Theory of Electromagnetic Beams

Synthesis Lectures on Engineering, Science, and Technology

Each book in the series is written by a well known expert in the field. Most titles cover subjects such as professional development, education, and study skills, as well as basic introductory undergraduate material and other topics appropriate for a broader and less technical audience. In addition, the series includes several titles written on very specific topics not covered elsewhere in the Synthesis Digital Library.

Theory of Electromagnetic Beams
John Lekner
2020

The Big Picture: The Universe in Five S.T.E.P.S.
John Beaver
2020

Relativistic Classical Mechanics and Electrodynamics
Martin Land and Lawrence P. Horwitz
2019

Generating Functions in Engineering and the Applied Sciences
Rajan Chattamvelli and Ramalingam Shanmugam
2019

Transformative Teaching: A Collection of Stories of Engineering Faculty's Pedagogical Journeys
Nadia Kellam, Brooke Coley, and Audrey Boklage
2019

Ancient Hindu Science: Its Transmission and Impact on World Cultures
Alok Kumar
2019

Value Rational Engineering
Shuichi Fukuda
2018

Strategic Cost Fundamentals: for Designers, Engineers, Technologists, Estimators, Project Managers, and Financial Analysts
Robert C. Creese
2018

Concise Introduction to Cement Chemistry and Manufacturing
Tadele Assefa Aragaw
2018

Data Mining and Market Intelligence: Implications for Decision Making
Mustapha Akinkunmi
2018

Empowering Professional Teaching in Engineering: Sustaining the Scholarship of Teaching
John Heywood
2018

The Human Side of Engineering
John Heywood
2017

Geometric Programming for Design Equation Development and Cost/Profit Optimization (with illustrative case study problems and solutions), Third Edition
Robert C. Creese
2016

Engineering Principles in Everyday Life for Non-Engineers
Saeed Benjamin Niku
2016

A, B, See... in 3D: A Workbook to Improve 3-D Visualization Skills
Dan G. Dimitriu
2015

The Captains of Energy: Systems Dynamics from an Energy Perspective
Vincent C. Prantil and Timothy Decker
2015

Lying by Approximation: The Truth about Finite Element Analysis
Vincent C. Prantil, Christopher Papadopoulos, and Paul D. Gessler
2013

Survive and Thrive: A Guide for Untenured Faculty
Wendy C. Crone
2010

Geometric Programming for Design and Cost Optimization (with Illustrative Case Study Problems and Solutions)
Robert C. Creese
2009

Style and Ethics of Communication in Science and Engineering
Jay D. Humphrey and Jeffrey W. Holmes
2008

Introduction to Engineering: A Starter's Guide with Hands-On Analog Multimedia Explorations
Lina J. Karam and Naji Mounsef
2008

Introduction to Engineering: A Starter's Guide with Hands-On Digital Multimedia and Robotics Explorations
Lina J. Karam and Naji Mounsef
2008

CAD/CAM of Sculptured Surfaces on Multi-Axis NC Machine: The DG/K-Based Approach
Stephen P. Radzevich
2008

Tensor Properties of Solids, Part Two: Transport Properties of Solids
Richard F. Tinder
2007

Tensor Properties of Solids, Part One: Equilibrium Tensor Properties of Solids
Richard F. Tinder
2007

Essentials of Applied Mathematics for Scientists and Engineers
Robert G. Watts
2007

Project Management for Engineering Design
Charles Lessard and Joseph Lessard
2007

Relativistic Flight Mechanics and Space Travel
Richard F. Tinder
2006

Theory of Electromagnetic Beams

John Lekner

ISBN: 978-3-031-00954-9   paperback
ISBN: 978-3-031-02082-7   ebook
ISBN: 978-3-031-00154-3   hardcover

DOI 10.1007/978-3-031-02082-7

A Publication in the Springer series
SYNTHESIS LECTURES ON ENGINEERING, SCIENCE, AND TECHNOLOGY

Lecture #39
Series ISSN
Print 2690-0300   Electronic 2690-0327

Theory of Electromagnetic Beams

John Lekner
Victoria University of Wellington, New Zealand

SYNTHESIS LECTURES ON ENGINEERING, SCIENCE, AND TECHNOLOGY #39

ABSTRACT

The theory of electromagnetic beams is presented in a simple and physical way, with all necessary mathematics explained in the text. The topics covered are in free-space classical electrodynamics, but contact is made with quantum theory in proofs that causal beams of various kinds can be viewed as superpositions of photons. This follows from explicit expressions for the energy, momentum and angular momentum per unit length for each type of beam. The properties of beams in the focal region, of special experimental and theoretical interest, are discussed in detail.

There are eight chapters: on Fundamentals, Beam-like solutions of the Helmholtz equation, Electromagnetic beams, Polarization, Chirality, Comparison of electromagnetic beams, a chapter on Sound beams and particle beams (to show the similarities to and differences from the vector electromagnetic beams), and a final chapter on Measures of focal extent. Ten Appendices cover mathematical or associated physical topics.

KEYWORDS

electromagnetic beams, Maxwell equations, classical electrodynamics, polarization, chirality *for Poppy and Dayton*

Contents

Preface

The aim of this short monograph is to present the theory of electromagnetic beams in a simple and physical way, from first principles. By which I mean from the Maxwell equations. The book can be short (just eight chapters) because the subject matter is restricted. It is based on free-space *classical* electrodynamics, and contact is made with quantum theory only in proofs that causal beams of various kinds can be viewed as superpositions of photons.

The subject matter is also restricted to theory: no experimental results are discussed, since the author does not feel qualified. (This is emphatically not because of lack of interest or respect: experiment is the ultimate arbiter in science.) Another restriction is to methodology: apart from a section on the paraxial approximation in Chapter 2, no approximations are made in the theory and in the examples. With the exception stated, only exact solutions feature.

The subject matter is further restricted to monochromatic electromagnetic beams which have finite energy, momentum and angular momentum per unit length of the beam. The properties of a beam in its focal region are of most experimental and theoretical interest, and are routinely discussed for each type of beam.

There are eight chapters: on Fundamentals, Beam-like solutions of the Helmholtz equation, Electromagnetic beams, Polarization, Chirality, Comparison of electromagnetic beams, a chapter on Sound beams and particle beams (to show the similarities to and differences from the vector electromagnetic beams), and a final somewhat technical chapter on Measures of focal extent. Ten Appendices cover mathematical or associated physical topics.

As a theorist I have naturally favored general results and have emphasized universal conservation and invariance properties. However, many examples of beams are made explicit in equations and in figures, to give the reader a feel for the physical meaning of the theory, and for the joy of the particular. Delight as well as devil is in the detail.

John Lekner
Wellington, New Zealand
February 2020

Acknowledgments

I am grateful to Rufus Boyack for many helpful comments.

John Lekner
February 2020

CHAPTER 1

Fundamentals

1.1 INTRODUCTION

The aim of this short monograph is to provide an introduction to the theory of monochromatic electromagnetic beams for scientists, especially those interested in tightly focused beams. We shall restrict the subject matter to those electromagnetic beams which satisfy the following criteria:

1. They satisfy Maxwell's equations.

2. They have finite energy, momentum and angular momentum per unit length of the beam.

3. They are causal, meaning that they do not contain backward-propagating elements far from the focal region.

(The one exception to Criterion 1 is Section 2.2 on the widely used paraxial approximation. Electromagnetic beams based on the paraxial approximation to solutions of the Helmholtz equation do not satisfy Maxwell's equations.)

The third criterion does not exclude the possibility (in fact it is a necessity, as we shall see) of backflow in the focal region, associated with the zeros of the complex beam wavefunction.

It is interesting that (to date) closed-form wavefunctions exist only for beams that contain backward-propagating parts, as discussed in Section 2.1. These backward-propagating parts can be made small by choice of parameters, and are experimentally necessary when mirrors are present, but are non-causal in free beam propagation.

This introductory Chapter summarizes the basic properties of electromagnetic beams.

1.2 MAXWELL EQUATIONS, VECTOR AND SCALAR POTENTIALS

We shall use Gaussian units, with some basic results repeated in SI units for completeness. Maxwell's free-space equations are, with $\partial_{ct} = c^{-1}\partial/\partial t$ (Jackson 1975 [7])

$$
\begin{aligned}
\nabla \cdot \boldsymbol{B} &= 0 & \nabla \cdot \boldsymbol{E} &= 0 \\
\nabla \times \boldsymbol{E} + \partial_{ct}\boldsymbol{B} &= 0 & \nabla \times \boldsymbol{B} - \partial_{ct}\boldsymbol{E} &= 0.
\end{aligned}
\tag{1.1}
$$

In SI units the Maxwell equations, in regions free of charge or current, read (Griffiths 1981 [6], Zangwill 2013 [20])

$$\nabla \cdot \boldsymbol{B} = 0 \qquad\qquad\qquad \nabla \cdot \boldsymbol{E} = 0$$
$$\nabla \times \boldsymbol{E} + \partial_t \boldsymbol{B} = 0 \qquad\qquad \nabla \times \boldsymbol{B} - c^{-2}\partial_t \boldsymbol{E} = 0. \qquad (1.1\text{SI})$$

The only differences lie in the positioning of the speed of light c. In Gaussian units the dimensionality of the electric and magnetic fields is the same, while in SI units $\boldsymbol{E}$ and $c\boldsymbol{B}$ have the same dimension. SI units use two defined constants ε_0, μ_0 as well as the speed of light, with $\varepsilon_0\mu_0 = c^{-2}$.

The energy, momentum and angular momentum densities of an electromagnetic field, in free space and in Gaussian units, are (Jackson 1975 [7])

$$u(\boldsymbol{r},t) = \frac{1}{8\pi}(E^2 + B^2), \qquad \boldsymbol{p}(\boldsymbol{r},t) = \frac{1}{4\pi c}\boldsymbol{E} \times \boldsymbol{B}, \qquad \boldsymbol{j}(\boldsymbol{r},t) = \boldsymbol{r} \times \boldsymbol{p}(\boldsymbol{r},t). \qquad (1.2)$$

The corresponding expressions in SI units read (Griffiths 1981 [6], Zangwill 2013 [20])

$$u(\boldsymbol{r},t) = \frac{1}{2}\left(\varepsilon_0 E^2 + \frac{1}{\mu_0}B^2\right), \qquad \boldsymbol{p}(\boldsymbol{r},t) = \frac{1}{\mu_0}\boldsymbol{E} \times \boldsymbol{B}, \qquad \boldsymbol{j}(\boldsymbol{r},t) = \boldsymbol{r} \times \boldsymbol{p}(\boldsymbol{r},t).$$
$$(1.2\text{SI})$$

We shall first restate the well-known result of electromagnetic theory, that all free-space solutions of Maxwell's equations can be expressed in terms of solutions of the wave equation. Electric and magnetic fields can be expressed in terms of the vector potential $\boldsymbol{A}(\boldsymbol{r},t)$ and scalar potential $V(\boldsymbol{r},t)$ via

$$\boldsymbol{E} = -\nabla V - \partial_{ct}\boldsymbol{A}, \qquad \boldsymbol{B} = \nabla \times \boldsymbol{A}. \qquad (1.3)$$

With these substitutions the source-free Maxwell equations $\nabla \cdot \boldsymbol{B} = 0$, $\nabla \times \boldsymbol{E} + \partial_{ct}\boldsymbol{B} = 0$ are satisfied automatically. If further $\boldsymbol{A}$ and V satisfy the Lorenz condition $\nabla \cdot \boldsymbol{A} + \partial_{ct}V = 0$, substitution of (1.3) into Maxwell's free space equations (of which the curl equations couple $\boldsymbol{E}$ and $\boldsymbol{B}$), decouples the vector and the scalar potentials:

$$\nabla^2\boldsymbol{A} - \partial_{ct}^2\boldsymbol{A} = 0, \qquad \nabla^2 V - \partial_{ct}^2 V = 0. \qquad (1.4)$$

For monochromatic beams of angular frequency $\omega = ck$ the time dependence is in the factor $e^{-i\omega t}$ in the case of complex field amplitudes, and in either $\cos\omega t$ or $\sin\omega t$ for real fields. In all cases the wave equations in (1.4) become vector or scalar Helmholtz equations of the form

$$\nabla^2\psi + k^2\psi = 0. \qquad (1.5)$$

When we are dealing with *complex field amplitudes*, and the corresponding complex potentials, the Lorenz condition $\nabla \cdot \boldsymbol{A} + \partial_{ct}V = 0$ becomes

$$\nabla \cdot \boldsymbol{A} - ikV = 0, \qquad \text{so} \quad V = (ik)^{-1}\nabla \cdot \boldsymbol{A}. \qquad (1.6)$$

Hence, for monochromatic complex fields and with the Lorenz condition satisfied, both the electric and the magnetic complex field amplitudes are given in terms of the complex vector potential $A(r)$:

$$E(r) = \frac{i}{k}[\nabla(\nabla \cdot A) + k^2 A], \qquad B(r) = \nabla \times A. \tag{1.7}$$

It is understood, in the use of complex amplitudes, that the real fields are obtained by taking the real or imaginary parts of the corresponding amplitude times $e^{-i\omega t}$. For example, $E(r,t)$ may be taken as either of the two following expressions:

$$Re\left\{E(r)e^{-i\omega t}\right\} = Re\left\{[E_r(r) + i\,E_i(r)]e^{-i\omega t}\right\} = E_r(r)\cos\omega t + E_i(r)\sin\omega t \tag{1.8}$$

$$Im\left\{E(r)e^{-i\omega t}\right\} = Im\left\{[E_r(r) + i\,E_i(r)]e^{-i\omega t}\right\} = E_i(r)\cos\omega t - E_r(r)\sin\omega t. \tag{1.9}$$

These fields differ only in time translation (by a quarter period): setting $\omega t \to \omega t - \pi/2$ morphs (1.8) into (1.9).

1.3   CONSERVATION LAWS, BEAM INVARIANTS

The energy, momentum and angular momentum densities of an electromagnetic field in free space are, in Gaussian units and with *real fields* $E(r,t)$, $B(r,t)$, given in (1.2).

The average of $u(r,t)$ over one period $2\pi/\omega$ is, in terms of the *complex field amplitudes* $E(r)$, $B(r)$,

$$\bar{u}(r) = \frac{1}{16\pi}\left\{E(r)\cdot E^*(r) + B(r)\cdot B^*(r)\right\} = \frac{1}{16\pi}\left\{E_r^2 + E_i^2 + B_r^2 + B_i^2\right\}. \tag{1.10}$$

Likewise the cycle-averaged momentum density is

$$\bar{p}(r) = \frac{1}{16\pi c}\left[E(r)\times B^*(r) + E^*(r)\times B(r)\right] = \frac{1}{8\pi c}\left[E_r \times B_r + E_i \times B_i\right]. \tag{1.11}$$

The conservation of energy equation, $\nabla \cdot S + \partial_t u = 0$, where $S = c^2 p$ is the energy flux density, has the cycle-average

$$\nabla \cdot \bar{p} = \partial_x \bar{p}_x + \partial_y \bar{p}_y + \partial_z \bar{p}_z = 0. \tag{1.12}$$

Applying $\int d^2 r = \int_{-\infty}^{\infty} dx \int_{-\infty}^{\infty} dy = \int_0^{\infty} d\rho\rho \int_0^{2\pi} d\phi$ to (1.12) gives, for transversely finite beams propagating in the z direction (Lekner 2004 [12])

$$\partial_z \int d^2 r\, \bar{p}_z = 0, \qquad \text{or} \qquad P_z' = \int d^2 r\, \bar{p}_z = \text{constant}. \tag{1.13}$$

We use the notation P_z', since $dP_z = P_z' dz$ is total z-component momentum contained in a transverse slice of the beam, of thickness dz. Equation (1.13) states that the momentum content per unit length, along the direction of net propagation of the beam, is an *invariant*. Note that the invariance of the *momentum* content per unit length is derived from the conservation of *energy* (the energy flux density is proportional to the momentum density).

The conservation of momentum equation is expressed in terms of the stress (or momentum flux density) tensor,

$$\partial_t p_i + \sum_j \partial_j \tau_{ij} = 0, \qquad \tau_{ij} = \frac{1}{4\pi}\left[\frac{1}{2}(E^2 + B^2)\delta_{ij} - E_i E_j - B_i B_j\right]. \tag{1.14}$$

(We use the sign of Barnett (2002) [2] in defining the stress tensor, so as to keep the same form for momentum conservation as we had for energy conservation, $\partial_t u + \nabla \cdot S = 0$. The stress tensor is not unique; we have chosen a symmetric, canonical form.) Taking the cycle average gives $\sum_j \partial_j \bar{\tau}_{ij} = 0$, and operating with $\int d^2 r$ gives $\partial_z \int d^2 r \bar{\tau}_{iz} = 0$ $(i = x, y, z)$. Thus momentum conservation leads to three invariants (Lekner 2004 [12])

$$\begin{aligned} T'_{xz} &= \int d^2 r \bar{\tau}_{xz} = -\frac{1}{4\pi}\int d^2 r \left[\overline{E_x E_z} + \overline{B_x B_z}\right] \\ T'_{yz} &= \int d^2 r \bar{\tau}_{yz} = -\frac{1}{4\pi}\int d^2 r \left[\overline{E_y E_z} + \overline{B_y B_z}\right] \\ T'_{zz} &= \int d^2 r \bar{\tau}_{zz} = \frac{1}{8\pi}\int d^2 r \left[\overline{E_x^2 + E_y^2 - E_z^2} + \overline{B_x^2 + B_y^2 - B_z^2}\right]. \end{aligned} \tag{1.15}$$

Three more invariants follow from the conservation of angular momentum,

$$\partial_t j_i + \sum_\ell \partial_\ell \mu_{\ell i} = 0, \qquad \mu_{\ell i} = \sum_j \sum_k \varepsilon_{ijk} x_j \tau_{k\ell}. \tag{1.16}$$

The angular momentum flux density tensor $\mu_{\ell i}$ is defined in terms of the momentum flux density tensor τ_{ij} (Barnett 2002 [2]); ε_{ijk} is the Levi–Civita symbol defined by $\varepsilon_{123} = 1$, $\varepsilon_{132} = -1$, etc. (even and odd permutations of 123, respectively, equal to $+1$, -1).

The angular momentum invariants are (Lekner 2004 [12])

$$\begin{aligned} M'_{zx} &= \int d^2 r \bar{\mu}_{zx} = \int d^2 r [y\bar{\tau}_{zz} - z\bar{\tau}_{yz}] \\ M'_{zy} &= \int d^2 r \bar{\mu}_{zy} = \int d^2 r [z\bar{\tau}_{xz} - x\bar{\tau}_{zz}] \\ M'_{zz} &= \int d^2 r \bar{\mu}_{zz} = \int d^2 r [x\bar{\tau}_{yz} - y\bar{\tau}_{xz}]. \end{aligned} \tag{1.17}$$

There are seven universal invariants of electromagnetic beams, arising from the conservation of energy, momentum and angular momentum. Perhaps surprisingly, the energy per unit length of the beam, $U' = \int d^2 r \bar{u}$, is not always an invariant, although it is constant for the types of generalized Bessel beams discussed in Lekner (2004b) [13], as is $J'_z = \int d^2 r \bar{j}_z$. These are the causal beams used throughout this book, so there are nine quantities associated with monochromatic causal beams, P'_z, U', J'_z, T'_{ij}, M'_{ij}, which are constant along the length of the beam. The invariants of particular types of beams will be discussed when these are introduced in Chapter 3.

The invariants for sound beams and quantum particle beams also correspond to conservation laws (Chapter 7). In both cases they originate from the conservation of particles (the continuity equation) and conservation of momentum and of angular momentum.

1.4   ENERGY-MOMENTUM INEQUALITY

In this book we choose the net direction of propagation of an electromagnetic beam to be along the z axis. We now show that $u > cp_z$, $U' > cP'_z$ for all transversely localized electromagnetic beams. (The other components of the *total* momentum are zero: beams converge onto their focal region or diverge from it, as we shall see, so they do have transverse momentum densities, but these are equal and opposite in parts of the beam symmetrically placed relative to the propagation axis.) The proof that $u > cp_z$ is elementary. Consider the energy and momentum densities $u(\mathbf{r}, t)$ and $p_z(\mathbf{r}, t)$. From (1.2) we have

$$
\begin{aligned}
8\pi \, (u - cp_z) &= \mathbf{E}^2 + \mathbf{B}^2 - 2(\mathbf{E} \times \mathbf{B})_z \\
&= E_x^2 + E_y^2 + E_z^2 + B_x^2 + B_y^2 + B_z^2 - 2\left(E_x B_y - E_y B_x\right) \qquad (1.18) \\
&= \left(E_x - B_y\right)^2 + \left(E_y + B_x\right)^2 + E_z^2 + B_z^2 \geq 0.
\end{aligned}
$$

Equality would require $E_z = 0 = B_z$ (purely transverse fields) and $E_x = B_y$ and $E_y = -B_x$. The divergence equations in (1.1) then give

$$
-\partial_x E_y + \partial_y E_x = 0 \qquad \text{and} \qquad \partial_x E_x + \partial_y E_y = 0. \qquad (1.19)
$$

Thus E_x and $-E_y$ would be a Cauchy–Riemann pair in the variables x and y, and satisfy

$$
\left(\partial_x^2 + \partial_y^2\right) E_x = 0, \qquad \left(\partial_x^2 + \partial_y^2\right) E_y = 0. \qquad (1.20)
$$

Such harmonic functions cannot have a maximum except at the boundary of their domain, and thus cannot be localized in x and y (for any z and t). For transversely localized electromagnetic beams we therefore always have the total energy per unit length of the beam greater than c times the total momentum per unit length of the beam,

$$
U' > cP'_z. \qquad (1.21)
$$

The analogous inequality linking total energy and the longitudinal component of total momentum, $U > cP_z$, holds for localized electromagnetic pulses (Lekner 2018 [16]). It also follows from (1.18). An alternative route to the inequality $U > cP_z$ is via the Fourier analysis of wavepackets (their construction by superposition of plane waves); see Zangwill (2013) [20], Section 16.5, and Problem 16.17.

We can show also that $u^2 - c^2 p^2$ is a Lorentz invariant, non-negative at all space-time points (u and $\boldsymbol{p}$ are the energy and momentum densities). This means that, for a particular solution of Maxwell's equations, the value of $u^2 - c^2 p^2$ is the same at corresponding space-time points in all inertial frames. Consider the squares of the volume densities, $u^2(\boldsymbol{r}, t)$ and $p^2(\boldsymbol{r}, t)$. From (1.2) we have (Lekner 2016 [16], Equation 19.66)

$$
\begin{aligned}
(8\pi)^2 \left(u^2 - c^2 p^2\right) &= \left(E^2 + B^2\right)^2 - 4\left(\boldsymbol{E} \times \boldsymbol{B}\right)^2 \\
&= \left(E^2 + B^2\right)^2 - 4E^2 B^2 + 4(\boldsymbol{E} \cdot \boldsymbol{B})^2 \\
&= \left(E^2 - B^2\right)^2 + 4(\boldsymbol{E} \cdot \boldsymbol{B})^2.
\end{aligned} \tag{1.22}
$$

Hence $u^2 - c^2 p^2$ is everywhere non-negative, and further it is a Lorentz invariant, since $E^2 - B^2$ and $\boldsymbol{E} \cdot \boldsymbol{B}$ are Lorentz invariants (see for example Zangwill 2013 [20], Section 22.7.2).

1.5   NON-EXISTENCE THEOREMS

In textbooks a light beam is usually represented by a plane wave, with $\boldsymbol{E}$, $\boldsymbol{B}$ and the propagation vector $\boldsymbol{k}$ everywhere mutually perpendicular. This 'beam' can be everywhere linearly polarized in the same direction, or everywhere circularly polarized in the same plane, and its energy is everywhere transported in a fixed direction at the speed of light. It has been shown in Lekner 2003 [11] that *none* of these properties can hold for a transversely finite beam.

We shall just state the theorems here, and give complete proofs in Appendix 4A, at the end of the chapter on Polarization.

(i) TEM beams ($\boldsymbol{E}$ and $\boldsymbol{B}$ both transverse to the propagation direction) do not exist.

(ii) Beams of fixed linear polarization do not exist.

(iii) Beams which are everywhere circularly polarized in the same plane do not exist.

(iv) Beams or pulses within which the energy velocity (Lekner 2003 [11]) is everywhere in the same direction and of magnitude c do not exist.

1.6   ZEROS OF THE BEAM WAVEFUNCTION IN THE FOCAL REGION

The solutions $\psi(\boldsymbol{r})$ of the Helmholtz equation are, in general, complex functions of position, $\psi = \psi_r + i\psi_i$. The real and imaginary parts ψ_r and ψ_i are (in free space) smooth functions of position. The functions ψ_r and ψ_i are zero on surfaces S_r and S_i, and where these surfaces meet (on curves C in space) both ψ_r and ψ_i are zero. If we write

$$
\psi(\boldsymbol{r}) = M(\boldsymbol{r})e^{iP(\boldsymbol{r})} = \left[\psi_r^2 + \psi_i^2\right]^{\frac{1}{2}} \exp\left(i \arctan \frac{\psi_i}{\psi_r}\right) \tag{1.23}
$$

we see that, on any such curve C, the modulus $M(r)$ is zero, and the phase $P(r)$ is indeterminate. Nye and Berry (1974) [17] called these curves wave dislocations; Chapter 5 of Nye's 1999 book [19] gives illustrations of such phase singularities.

Lekner (2016) [15] has given a topological argument for the existence of zeros of ψ in the focal region, on the assumption that the isophase surfaces intersect the focal plane. At the zeros of ψ the phase can be any real number, but integer multiples of π are excluded, as explained below. The original conjecture (Lekner 2016 [15], Section 20.1.4) was that the zeros all lie in the focal plane, which is a plane of symmetry for an ideal beam. However, much earlier Carter (1973) [5] had seen 'anomalies' in the focal region, some of which lay away from the focal plane. Later Karman et al. (1997, 1998) [9, 10] showed experimentally and theoretically that phase singularities in the focal region can be created and annihilated, and further that they can move off the focal plane. Berry (1998) [3] and Nye (1998) [18] have discussed this phenomenon. Andrejic and Lekner (2017) [1] examined its effect on the polarization properties of circularly and linearly polarized beams. The zeros of ψ are in the focal region in all the cases studied, but are not always confined to the focal plane.

A general argument for the existence of zeros thus must allow for the possibility of these zeros moving off the focal plane. For simplicity we shall assume here that they do lie in the focal plane, which can be taken as the $z = 0$ plane. We can also take the phase of ψ to be zero at the origin. Then the isophase surfaces correspond to negative $P(r)$ for $z < 0$ and positive $P(r)$ for $z > 0$. The isophase surfaces are concave toward the origin, since a physical beam is converging toward the focal region for $z < 0$ and diverging from it for $z > 0$. The surfaces $P = -n\pi$ and $P = n\pi$ can meet where ψ is not zero, since the phase difference is an integer (n) multiple of 2π. The other isophase surfaces can only meet on the focal plane if on it there exist curves where ψ is zero. On such curves (circles in the focal plane, in the simplest case) the phase surfaces $P = -\pi/2$ and $P = +\pi/2$ can meet, for example. The surfaces with $0 < |P| < \pi$ meet on the first zero curve, $\pi < |P| < 2\pi$ meet on the next, and so on. Figure 1.1 illustrates the phenomenon for a particular beam wavefunction. Because of the topological nature of the above argument, we expect the zeros to persist even when the beam is perturbed (for example, focused by an imperfect lens or mirror). The focal plane would then be distorted to a nearly-planar surface, and the circles of zeros to approximately circular closed curves, where the perturbed phase surfaces $\pm P$ meet.

One counter-example to the above conjecture (of the universality of rings of zeros in the focal region) would seem to be separable spheroidal beams, for which $\psi(\xi, \eta, \phi) = R(\xi)S(\eta)e^{im\phi}$ with $\rho = b\left[(\xi^2 + 1)\left(1 - \eta^2\right)\right]^{\frac{1}{2}}$, $z = b\xi\eta$. The focal plane $z = 0$ corresponds to $\xi = 0$ for $\rho \leq b$ and $\eta = 0$ for $\rho \geq b$. Thus if $S(\eta)$ is zero for $\eta = 0$, $\psi = 0$ for $\rho \geq b$ in the focal plane, and the $-P$ and $+P$ isophase surfaces can meet anywhere on the focal plane outside of the central disk $\rho \leq b$. However, such spheroidal wavefunctions have been shown to be non-physical (Boyack and Lekner 2011 [4]).

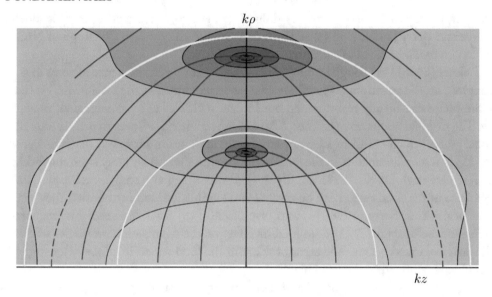

Figure 1.1: The proto-beam wavefunction $\psi_0 (\rho, z)$ (Lekner 2016 [15] and Section 2.5) in the focal region: contours of constant modulus (logarithmic scale), and isophase surfaces (red and yellow). The three-dimensional picture is obtained by rotating the figure about the beam axis (the horizontal axis). The phase is chosen to be zero at the origin, and isophase contours are shown at intervals of $\pi/3$. The π, $-\pi$, and 2π, -2π phase surfaces (yellow) are hemispherical. Note that the isophase surfaces (apart from those which are integral multiples of π) meet on the zeros of ψ_0, which lie in the focal plane. The extent of the focal region shown is in $k\,|z| \le 8$, $k\rho \le 8$.

Figure 1.2 shows detail of Figure 1.1 in the neighborhood of the first zero of the proto-beam. The surfaces of constant modulus tend to tori of elliptical cross-section as the zero is approached. If $(k\rho)_n$ is a zero of $\psi_0 (\rho, 0) = 2J_1(k\rho)/k\rho$, the curves of constant modulus in a given plane through the beam axis (the z axis) tend to

$$\frac{z^2}{a^2} + \frac{(\rho - \rho_n)^2}{b^2} = 1, \qquad \text{with} \qquad \frac{a}{b} = \left| \frac{x^2 J_0(x)}{\sin x \ - x\cos x} \right|, \qquad x = (k\rho)_n. \qquad (1.24)$$

For the first zero at $(k\rho)_1 \approx 3.8317$, the ratio of the semi-axes is approximately 2.55.

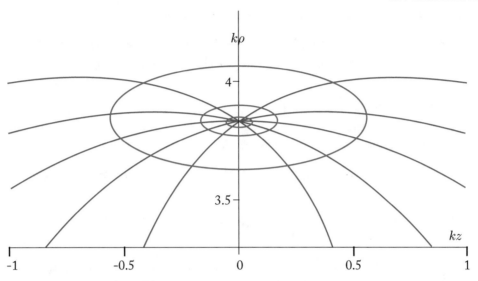

Figure 1.2: Detail of the proto-beam wavefunction $\psi_0\left(\rho, z\right)$ in the neighborhood of the first zero: contours of constant modulus (logarithmic scale, blue), and isophase surfaces (red). As in Figure 1.1, except that the isophase contours are shown at intervals of $\pi/6$. Near the zero the surfaces of constant modulus tend to tori of elliptical cross-section, as discussed in the text.

1.7 REFERENCES

[1] Andrejic, P. and Lekner, J. 2017. Topology of phase and polarization singularities in focal regions, *Journal of Optics*, 19(105609):8. DOI: 10.1088/2040-8986/aa895d. 7

[2] Barnett, S. M. 2002. Optical angular momentum flux, *Journal of Optics B: Quantum Semiclassical Optics*, 4(S1–S10). DOI: 10.1088/1464-4266/4/2/361. 4

[3] Berry, M. V. 1998. Wave dislocation reactions in non-paraxial Gaussian beams, *Journal of Modern Optics*, 45:1845–1858. DOI: 10.1080/09500349808231706. 7

[4] Boyack, R. and Lekner, J. 2011. Non-existence of separable spheroidal beams, *Journal of Optics*, 13(085701):3. DOI: 10.1088/2040-8978/13/8/085701. 7

[5] Carter, W. H. 1973. Anomalies in the field of a Gaussian beam near focus, *Optics Communications*, 64:491–495. DOI: 10.1016/0030-4018(73)90012-6. 7

[6] Griffiths, D. 1981. *Introduction to Electrodynamics*, Prentice Hall. DOI: 10.1017/9781108333511. 2

[7] Jackson, J. D. 1975. *Classical Electrodynamics*, 2nd ed., Wiley. DOI: 10.1063/1.3057859. 1, 2

[8] Jones, D. S. 1964. *The Theory of Electromagnetism*, Pergamon.

[9] Karman, G. P., Beijersbergen, M. W., van Duijl, A., and Woerdman, J. P. 1997. Creation and annihilation of phase singularities in a focal field, *Optics Letters*, 22:1503–1505. DOI: 10.1364/ol.22.001503. 7

[10] Karman, G. P., Beijersbergen, M. W., van Duijl, A., Bouwmeester, D., and Woerdman, J. P. 1998. Airy pattern reorganization and subwavelength structure in a focus, *Journal of the Optical Society of America A*, 15:884–899. DOI: 10.1364/josaa.15.000884. 7

[11] Lekner, J. 2003. Polarization of tightly focused laser beams, *Journal of Optics A: Pure and Applied Optics*, 5:6–14. DOI: 10.1088/1464-4258/5/1/302. 6

[12] Lekner, J. 2004. Invariants of electromagnetic beams, *Journal of Optics A: Pure and Applied Optics*, 6:204–209. DOI: 10.1088/1464-4258/6/2/008. 3, 4

[13] Lekner, J. 2004b. Invariants of three types of generalized Bessel beams, *Journal of Optics A: Pure and Applied Optics*, 6:837–843. DOI: 10.1088/1464-4258/6/9/004. 4

[14] Lekner, J. 2016. *Theory of Reflection*, 2nd ed., Springer International. DOI: 10.1007/978-3-319-23627-8.

[15] Lekner, J. 2016. Tight focusing of light beams: A set of exact solutions, *Proc. Royal Society A*, 472(20160538):17. DOI: 10.1098/rspa.2016.0538. 7, 8

[16] Lekner, J. 2018. *Theory of Electromagnetic Pulses*, IOP Concise Physics, Bristol. DOI: 10.1088/978-1-6432-7022-7. 5, 6

[17] Nye, J. F. and Berry, M. V. 1974. Dislocations in wave trains, *Proc. Royal Society of London A*, 336:165–190. DOI: 10.1098/rspa.1974.0012. 7

[18] Nye, J. F. 1998. Unfolding of higher-order wave dislocations, *Journal of Optics Society of America*, A(15):1132–1138. DOI: 10.1364/josaa.15.001132. 7

[19] Nye, J. F. 1999. *Natural Focusing and the Fine Structure of Light*, Institute of Physics Publishing, Bristol. 7

[20] Zangwill, A. 2013. *Modern Electrodynamics*, Cambridge University Press. DOI: 10.1017/cbo9781139034777. 2, 5, 6

CHAPTER 2

Beam-Like Solutions of the Helmholtz Equation

2.1 INTRODUCTION

We saw in Section 1.2 that for monochromatic beams of angular frequency $\omega = ck$ the time dependence is in the factor $e^{-i\omega t}$ in the case of complex field amplitudes, and either $\cos \omega t$ or $\sin \omega t$ for real fields. In all cases the wave equations in (1.4) become vector or scalar Helmholtz equations of the form

$$\nabla^2 \psi + k^2 \psi = 0. \tag{2.1}$$

The Helmholtz equation is a partial differential equation, with an uncountable infinity of solutions. Textbook examples are the plane waves $e^{i\mathbf{k}\cdot\mathbf{r}}$ and spherically diverging or converging waves $r^{-1}e^{ikr}$, $r^{-1}e^{-ikr}$ ($k^2 = k_x^2 + k_y^2 + k_z^2$). These textbook solutions of (2.1) are not bounded in space. Physical beams are localized transversely to the direction of propagation, in contradistinction to the textbook solutions.

Deschamps (1971) [9] noted that a complex shift in the coordinates applied to a solution of (2.1) is also a solution, since the Laplacian is translationally invariant in x, y or z. We complex-shift along the propagating direction (the z axis, in this book), by setting $z \to z - ib$ in the spherical wave $r^{-1}e^{ikr}$. The resulting wavefunction is

$$\psi = \frac{e^{ikR}}{R}, \qquad R^2 = x^2 + y^2 + (z - ib)^2 = \rho^2 + (z - ib)^2. \tag{2.2}$$

This solution is singular on the circle $\{\rho = b, z = 0\}$, and so cannot represent a physical beam. One can regularize by subtracting the complex-shifted spherically converging wave $\exp(-ikR)/R$ (Sheppard and Saghati 1998 [26]), to obtain

$$\psi = \frac{\sin kR}{kR} = j_0(kR). \tag{2.3}$$

Ulanowski and Ludlow (2000) [28] generalized this to the set formed from products of spherical Bessel functions and Legendre polynomials, both functions of complex-shifted spatial coordinates:

$$\psi_{\ell m} = j_\ell(kR) P_\ell^m \left(\frac{z - ib}{R} \right) e^{im\phi}. \tag{2.4}$$

Each of the set in (2.4) is an exact solution of the Helmholtz equation, but will not be used in this book because of two problems: first the divergence of some integrals, for example the normalization integral $\int d^2r|\psi_{00}|^2$ (Lekner 2001 [14]). Second, there are backward-propagating components associated with the terms proportional to e^{-ikR}. Such backward propagation can exist in the presence of mirrors, but we are concerned with free-space propagation of beams originating in a single coherent source. The remainder of this chapter deals with beam wavefunctions that represent purely forward-propagating beams. We begin with some simple approximate solutions.

2.2   PARAXIAL APPROXIMATION OF THE HELMHOLTZ EQUATION

The name 'paraxial' originated in ray-tracing and refers to the ray angle to the optic axis being small. In wave optics 'paraxial' is used to describe an approximation which amounts to assuming that the dominant z-dependence of the beam lies in the e^{ikz} factor modulated by a slowly-varying function.

The Helmholtz Equation 2.1 is separable in cylindrical coordinates (ρ, ϕ, z), in which it reads

$$\left[\partial_\rho^2 + \frac{1}{\rho}\partial_\rho + \frac{1}{\rho^2}\partial_\phi^2 + \partial_z^2 + k^2\right]\psi = 0. \tag{2.5}$$

In this Section we consider Gaussian beams, solutions of the *paraxial equation*, in which one sets $\psi = e^{ikz}G$, and then neglects the term $\partial_z^2 G$ in the resulting equation for G (given below). For axially symmetric solutions we omit the azimuthal derivative, so the equation for G becomes $(\partial_\rho^2 + \rho^{-1}\partial_\rho + 2ik\partial_z + \partial_z^2)G = 0$. The paraxial equation has the ∂_z^2 term omitted:

$$\left(\partial_\rho^2 + \rho^{-1}\partial_\rho + 2ik\partial_z\right)G \approx 0. \tag{2.6}$$

The fundamental mode solution of (2.6) is (see for example Zangwill 2013 [31], Section 16.7)

$$\psi_G = e^{ikz}G = \frac{b}{b+iz}\exp\left\{ikz - \frac{k\rho^2}{2(b+iz)}\right\}. \tag{2.7}$$

Alternatively, we can write the Gaussian beam fundamental mode in the modulus times phase factor form:

$$\psi_G = \frac{b}{\sqrt{b^2+z^2}}\exp\left[\frac{-kb\rho^2}{2(b^2+z^2)}\right]\exp i\left\{kz - \arctan\left(\frac{z}{b}\right) + \frac{kz\rho^2}{2(b^2+z^2)}\right\}. \tag{2.8}$$

The arctangent term in the phase causes the phase of the beam to decrease by π relative to the plane wave phase kz as the beam passes through its focal region. This phase lag associated with focusing is universal for waves. It was first noted by Gouy in 1890 [11].

The length b is the Rayleigh or diffraction length: it gives the longitudinal extent of the beam focal region directly, as one can see from the modulus exponential in (2.8). The beam

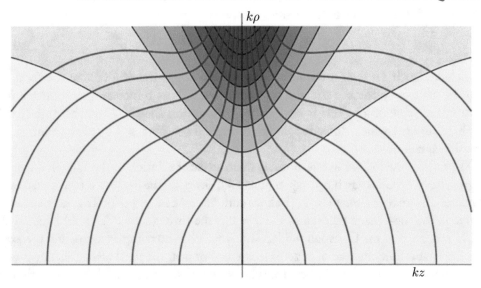

Figure 2.1: $\psi_G(\rho, z)$ in the focal region, plotted for $kb = 2$, for $k|z| \leq 9$, $k\rho \leq 9$. Shading indicates modulus of the wavefunction (logarithmic scale, lighter color indicates larger modulus). The isophase surfaces are shown in red at intervals of $\pi/2$. The phase is chosen to be zero at the origin. The three-dimensional picture is obtained by rotating the figure about the beam axis (the horizontal axis).

waist size, located at the focal plane (here $z = 0$) is obtained from the modulus in the focal plane, $\exp(-k\rho^2/2b)$. It is usually written as $w_0 = \sqrt{2b/k} = \sqrt{b\lambda/\pi}$. If we define the beam width $w(z)$ by setting the exponential in the modulus equal to $\exp\left[-\rho^2/w^2(z)\right]$, we get

$$w^2(z) = \frac{2(b^2 + z^2)}{kb}. \tag{2.9}$$

Thus $w = w_0$ at the waist, $w^2 = 2w_0^2$ at $z = \pm b$, and $w^2 \to 2z^2/kb$ when $|z| \gg b$. Away from the focal region the beam spreads as a cone of half-angle $\arctan\sqrt{2/kb}$. For $kb = 2$ and 6 this angle is 45° and 30°, respectively. Figure 2.1 shows the Gaussian $kb = 2$ fundamental mode. Note that the isophase surfaces meet at infinity, the only place where the wavefunction ψ_G is zero.

We need to consider the validity of the paraxial approximation which leads to the Gaussian beam solution. As noted in Lekner 2001 [14], the quantity $\psi_G^{-1}\nabla^2\psi_G$ should equal $-k^2$, but instead equals $-k^2$ times

$$1 + \frac{2}{k^2(b + iz)^2} - \frac{2\rho^2}{k(b + iz)^3} + \frac{\rho^4}{4(b + iz)^4}. \tag{2.10}$$

The errors are thus negligible in the regions where

$$k^2 \left(b^2 + z^2\right) \gg 1 \qquad \text{and} \qquad b^2 + z^2 \gg \rho^2. \tag{2.11}$$

We conclude that if kb is of order unity, the paraxial approximation fails in the focal region $|z| \leq b$. In the focal plane the paraxial approximation fails as ρ increases to approach b, irrespective of the value of kb. This is because the assumption that $\partial_z^2 G$ may be neglected leads to isophase surfaces which differ topologically from those obtained by exact solution of the Helmholtz equation.

We can ask the following questions: is there an exact solution of the Helmholtz equation that agrees with the Gaussian beam expression (2.7), either on the beam axis (the z axis) or in the focal plane? The answer is *no* to both (Lekner and Andrejic 2018 [20]). That no exact solution can agree in the focal plane (chosen as the $z = 0$ plane throughout this book) is noted above: from (2.8) we see that the Gaussian fundamental mode has no complex zeros, and the isophase surfaces meet at infinity instead of at zeros located on or near the focal plane (Section 1.6). That no exact solution can agree on the beam axis is a surprise, since a paraxial approximation might be expected to be good on and near the axis.

The Gaussian beam fundamental mode $\psi_G = \psi_{00}$ can be used to generate modes with azimuthal dependence. There are two classes of such beams, as illustrated below:

$$\psi_{10} = \frac{x}{b + iz} \psi_{00}, \qquad \psi_{11} = \frac{x}{b + iz} \frac{y}{b + iz} \psi_{00} \qquad \text{(Hermite-Gaussian)} \tag{2.12}$$

$$\psi_{10} \pm i \psi_{01} = \frac{x \pm iy}{b + iz} \psi_{00} = \frac{\rho e^{\pm i\phi}}{b + iz} \psi_{00} \qquad \text{(Laguerre–Gaussian)}. \tag{2.13}$$

2.3   SOLUTIONS IN CYLINDRICAL COORDINATES

In cylindrical polar coordinates (ρ, ϕ, z), with $\rho = \left(x^2 + y^2\right)^{\frac{1}{2}}$ the distance from the z-axis and ϕ the azimuthal angle, the Helmholtz equation is $(\partial_\rho^2 + \rho^{-1} \partial_\rho + \rho^{-2} \partial_\phi^2 + \partial_z^2 + k^2)\psi = 0$. This is satisfied by $J_m (\kappa\rho) e^{im\phi} e^{iqz}$ if the transverse and longitudinal wavenumber components $\kappa = k_\rho$ and $q = k_z$ are constrained by $\kappa^2 + q^2 = k^2$. $J_m (\kappa\rho)$ is the regular Bessel function of order m. The general expression for monochromatic beams of frequency $\omega = ck$ and azimuthal winding number m contains a wavenumber weight function $f(k, \kappa)$:

$$\psi_m (r, k) = e^{im\phi} \int_0^k d\kappa \, f(k, \kappa) \, e^{iqz} J_m (\kappa\rho), \qquad q = \sqrt{k^2 - \kappa^2}. \tag{2.14}$$

These are causal solutions, by which we mean without backward propagation far from the focal region. (There is in general backflow associated with the zeros of the beam wavefunction; these zeros lie in the focal region, as discussed in Section 1.6.) The function $f(k, \kappa)$, in general complex, is subject only to the existence of (2.14) and of associated integrals, for example those which give the energy, momentum and angular momentum contained in a transverse slice of a beam constructed from $\psi(\rho, \phi, z)$. The form of (2.14) guarantees the absence of asymptotic backward propagation: the integrand contains the factor e^{iqz}, with $k \geq q \geq 0$. For ψ to be dimensionless, the dimension of $f(k, \kappa)$ is to be that of a length. The values of m are restricted to integers by the condition $\psi(\rho, \phi + 2\pi, z) = \psi(\rho, \phi, z)$.

As an example of the constraints on $f(k, \kappa)$, we may require the integral of $|\psi_m|^2$ over a section of the beam taken at constant z to be finite, as defined and evaluated in the following:

$$
\begin{aligned}
N' &= \int d^2r |\psi_m|^2 = \int_0^\infty d\rho\, \rho \int_0^{2\pi} d\phi |\psi_m|^2 \\
&= 2\pi \int_0^\infty d\rho\, \rho \int_0^k d\kappa\, f(k,\kappa)^* e^{-iqz} J_m(\kappa\rho) \int_0^k d\kappa'\, f(k,\kappa') e^{iq'z} J_m(\kappa'\rho).
\end{aligned}
\tag{2.15}
$$

The primed variables are related by $q' = \sqrt{k^2 - \kappa'^2}$. The notation N' follows that of Section 1.3: $dN = N' dz$ is the probability content in a slice dz of the beam, so $N' = dN/dz$.

Hankel's inversion formula (Watson 1944 [30], Section 14.4) may be written as

$$
\int_0^\infty d\rho\, \rho\, J_m(\kappa\rho)\, J_m(\kappa'\rho) = \kappa^{-1} \delta(\kappa - \kappa') \qquad (\kappa, \kappa' > 0).
\tag{2.16}
$$

Hence performing the integration over ρ selects $\kappa' = \kappa$, and we are left with

$$
N' = 2\pi \int_0^k d\kappa\, \kappa^{-1} f(k,\kappa)^* f(k,\kappa) = 2\pi \int_0^k d\kappa\, \kappa^{-1} |f(k,\kappa)|^2.
\tag{2.17}
$$

We see that the function $f(k, \kappa)$ cannot be chosen arbitrarily. For example, logarithmic divergence would result from a nonzero value of $f(k, 0)$, making a finite norm impossible. The result (2.17) shows that the norm of generalized Bessel beams defined in (2.14) is independent of z: the value of $\int d^2r |\psi_m|^2$ is the same along the entire extent of the beam (scattering and absorption being neglected).

2.4   BATEMAN'S INTEGRAL SOLUTION

Bateman (1904) [5] obtained a general solution of the wave equation in integral form. For solutions with axial symmetry (independent of the azimuthal angle ϕ) the Bateman solution is, with $F(u, v)$ an arbitrary twice-differentiable function,

$$
\Psi(\rho, z, t) = \frac{1}{2\pi} \int_{-\pi}^{\pi} d\theta\, F(z + i\rho\cos\theta, ct + \rho\sin\theta).
\tag{2.18}
$$

We can adapt this to find the general non-singular solution of the Helmholtz equation 2.1 which is independent of the azimuthal angle. For time-dependence $e^{-i\omega t} = e^{-ikct}$, the function F must take the form

$$
F(z + i\rho\cos\theta, ct + \rho\sin\theta) = g(z + i\rho\cos\theta) e^{-ik(ct + \rho\sin\theta)}.
\tag{2.19}
$$

The spatial part of Ψ in (2.18) then becomes

$$\psi(\rho, z) = \frac{1}{2\pi} \int_0^{2\pi} d\theta\, g(z + i\rho\cos\theta) e^{-ik\rho\sin\theta}. \tag{2.20}$$

We can verify that this is a solution of the Helmholtz equation as follows. Let $G(\rho, z, \theta) = g(z + i\rho\cos\theta)e^{-ik\rho\sin\theta}$. A short calculation shows that $(\nabla^2 + k^2)G = -\rho^{-2}\partial_\theta^2 G$ and so $2\pi(\nabla^2 + k^2)\psi = [\rho^{-2}\partial_\theta G]_0 - [\rho^{-2}\partial_\theta G]_{2\pi} = 0$. Thus the expression (2.20) is the most general form of the non-singular scalar wavefunction corresponding to axially symmetric monochromatic beams. (We wrote 'non-singular' because Bateman also gave a solution which, when adapted to monochromatic beams, is singular on the beam axis.)

Note that on the beam axis ($\rho = 0$) we get $\psi(0, z) = g(z)$. Therefore the amplitude function g in (2.20) is given by the axial value of the beam wavefunction.

There is a one-to-one correspondence between (2.20) and the $m = 0$ generalized Bessel beam solution (2.14). Since $\kappa^2 + q^2 = k^2$ we can write $\psi_0(\boldsymbol{r})$ as an integral over q instead of over κ:

$$\psi_0(\boldsymbol{r}) = \int_0^k dq\, h(k, q)\, J_0\left(\rho\sqrt{k^2 - q^2}\right) e^{iqz}, \qquad h(k, q) = \frac{q}{\kappa} f(k, \kappa). \tag{2.21}$$

The zero-order Bessel function containing the square root can be rewritten by using Bessel's integral (Watson 1944 [30], Section 2.21), which transforms (2.21) into

$$\psi_0(\rho, z) = \frac{1}{2\pi} \int_0^{2\pi} d\theta e^{-ik\rho\sin\theta} \int_0^k dq\, h(k, q)\, e^{iq(z + i\rho\cos\theta)}. \tag{2.22}$$

Comparison of (2.20) and (2.22) shows that, for the $m = 0$ generalized Bessel beams, the amplitude function g is given by

$$g(z + i\rho\cos\theta) = \int_0^k dq\, h(k, q)\, e^{iq(z + i\rho\cos\theta)}. \tag{2.23}$$

The axial value of the beam wavefunction is thus equal to the finite Fourier transform of $h(k, q)$:

$$\psi(0, z) = g(z) = \int_0^k dq\, h(k, q)\, e^{iqz}. \tag{2.24}$$

As an example we set $h(k, q) = 2q/k^2$, corresponding to the proto-beam studied by Lekner (2016) [18]. The axial value of this beam is

$$\psi(0, z) = g(z) = 2(kz)^{-2}\left[(1 - ikz)e^{ikz} - 1\right]. \tag{2.25}$$

The beam wavefunction at all points in space is obtained from (2.20) on replacing z by $z + i\rho\cos\theta$ in (2.25). The proto-beam is the confluent limit of two families of beams, to be discussed in the next section.

2.5   EXACT BEAM WAVEFUNCTIONS

We shall first look at $m = 0$ wavefunctions in the form (2.14) or (2.21). The earliest of this type appears to that of Carter (1973) [8],

$$
\begin{aligned}
\psi_C(\rho, z) &= \frac{b/k}{1 - e^{-kb/2}} \int_0^k d\kappa \, \kappa \, e^{-b\kappa^2/2k + iqz} J_0(\kappa\rho) \\
&= \frac{b/k}{e^{kb/2} - 1} \int_0^k dq \, q \, e^{bq^2/2k + iqz} J_0(\kappa\rho).
\end{aligned}
\tag{2.26}
$$

The prefactor to the integral is chosen to normalize the beam wavefunction to unity at the origin. We shall compare the Carter beams to the family obtained from the proto-beam by a complex shift in z (Lekner 2016 [17])

$$
\begin{aligned}
\psi_b(\rho, z) &= \frac{b^2}{e^{kb}(kb - 1) + 1} \int_0^k d\kappa \, \kappa \, e^{qb + iqz} J_0(\kappa\rho) \\
&= \frac{b^2}{e^{kb}(kb - 1) + 1} \int_0^k dq \, q \, e^{qb + iqz} J_0(\kappa\rho).
\end{aligned}
\tag{2.27}
$$

As the dimensionless parameter kb tends to zero, the families (2.26) and (2.27) both tend to the proto-beam (shown in Figure 1.1),

$$
\psi_0(\rho, z) = \frac{2}{k^2} \int_0^k d\kappa \, \kappa \, e^{iqz} J_0(\kappa\rho) = \frac{2}{k^2} \int_0^k dq \, q \, e^{iqz} J_0(\kappa\rho).
\tag{2.28}
$$

The proto-beam is shown in Chapter 8 to be the most tightly focused of all possible beams, according to an intensity criterion. Explicit expressions for ψ_0 are given in Lekner 2016 [17], in terms of Lommel functions of two variables, and alternatively in terms of a series expansion in spherical Bessel functions and Legendre polynomials. These will be given in Appendix 2B, which gives more detail on the properties of the proto-beam. In Section 2.6 we discuss a generalization of the proto-beam to the family of beams (2.27) characterized by the length b as well as by the wavenumber k.

The proto-beam has all its complex zeros in the focal plane $z = 0$, in which $\psi_0(\rho, 0) = 2J_1(k\rho)/k\rho$. All the zeros of $\psi_b(\rho, z)$ also lie in the focal plane, for all values of kb. The behavior of the Carter beam family is topologically different, except at small kb (Berry 1998 [6], Nye 1998 [21], Andrejic and Lekner 2017 [1]). For kb less than about 4.9196 all the zeros lie in the focal plane, but for larger values of kb some move off-plane. The behavior is quite intricate: Berry (1998) [6] has identified three topological events, portrayed in Figures 3 and 4 of Andrejic and Lekner (2017) [1].

The wavefunctions ψ_C and ψ_b coalesce into the proto-beam ψ_0 as $kb \to 0$, but as noted some of the Carter zeros move off-plane at intermediate and large values of the parameter kb. It is interesting that the difference arises as the *focus is loosened* (kb small corresponds to tight focus, $kb = 0$ gives the tightest focus common to both families). Measures of extent of the focal region are discussed in Section 2.7. In this context we note two beam wavefunctions that are also contenders for the most tightly-focused cachet.

The first is due to Porras (1994) [24], who took the wavenumber weight function proportional to $\kappa J_0(\kappa a)$, with the length a chosen so that $J_0(ka) = 0$. For example, $ka \approx 2.4048$ gives the first zero of the Bessel function J_0. The Porras wavefunction, normalized to unity at the origin, is

$$\psi_{Porras}(\rho, z) = \frac{a}{k J_1(ka)} \int_0^k d\kappa \, \kappa \, J_0(\kappa a) J_0(\kappa \rho) e^{iqz} \qquad (J_0(ka) = 0). \qquad (2.29)$$

The reason for this choice of $f(k, \kappa)$ is that it gives the smallest possible (of all solutions of the Helmholtz equation) second moment $\langle \rho^2 \rangle$ in the focal plane, discussed in Section 2.7. In fact *only* by making $J_0(\kappa a)$ zero at the upper limit of the integral do we get a *finite* second moment. In the focal plane

$$\psi_{Porras}(\rho, 0) = a \frac{a J_0(k\rho) J_1(ka) - \rho J_1(k\rho) J_0(ka)}{[a^2 - \rho^2] J_1(ka)}$$

$$\left(= \frac{a^2 J_0(k\rho) J_1(ka)}{[a^2 - \rho^2] J_1(ka)} \quad \text{if} \quad J_0(ka) = 0 \right). \qquad (2.30)$$

When $J_0(ka) = 0$ the asymptotic form of (2.30) drops by one power of ρ, which is sufficient to give convergence in the evaluation of $\langle \rho^2 \rangle$. Any of the roots of $J_0(ka) = 0$ give a beam with finite second moment. The singularity in (2.30) at $\rho = a$ is removable, whether or not ka is chosen to be a zero of J_0. ψ_{Porras} is forward-propagating in the same sense as ψ_0 and ψ_b, by construction.

The second beam wavefunction is that of Philbin (2018) [23], and is constructed as the sum of a standing wave and of an integral over a standing wave which makes the total wave forward-propagating:

$$\psi_{Philbin}(\rho, z) = \frac{\sin kr}{kr} + \frac{i}{k} \int_0^k dq \, J_0(\kappa \rho) \sin qz . \qquad (2.31)$$

(A simple example of two standing waves together making a progressive wave is $\cos kz + i \sin kz = e^{ikz}$.) On the beam axis ($\rho = 0$) we have $r = |z|$ and

$$\psi_{Philbin}(0, z) = \frac{\sin k|z|}{k|z|} + \frac{i}{kz}(1 - \cos kz) = \frac{e^{ikz} - 1}{ikz} = \frac{2}{kz} \sin \frac{kz}{2} e^{ikz/2}. \qquad (2.32)$$

This is clearly forward-propagating. There are wavefunction zeros on the beam axis, at $|kz| = 2n\pi$, $n = 1, 2, \dots$. We note that the expressions in (2.32) are equal to $k^{-1} \int_0^k dq e^{iqz}$, so [compare (2.21)] we may write the Philbin beam in the compact form

$$\psi_{Philbin}(\rho, z) = \frac{1}{k} \int_0^k dq \, e^{iqz} J_0 \left(\rho \sqrt{k^2 - q^2} \right). \qquad (2.33)$$

In the focal plane this simplifies to the form evident in (2.31),

$$\frac{1}{k} \int_0^k dq \, J_0 \left(\rho \sqrt{k^2 - q^2} \right) = \frac{1}{k} \int_0^k d\kappa \, \kappa \frac{J_0(\kappa\rho)}{\sqrt{k^2 - \kappa^2}} = \frac{\sin k\rho}{k\rho}. \tag{2.34}$$

Both the Porras and the Philbin beams may be generalized to beam *families* by the complex displacement $z \to z - ib$ along the beam axis, which has the effect of replacing e^{iqz} by e^{iqz+qb} in the integrands of (2.29) and (2.33). Both families are characterized by the length k^{-1} and the dimensionless parameter kb, just as in the case of the family $\psi_b(\rho, z)$ defined in (2.27), to be discussed in the next section.

Unfortunately, the Philbin beam is not a viable physical beam because it does not decrease rapidly enough in the transverse direction. For example, from (2.17) the integral at constant z of the probability density over a section of a beam with wavefunction ψ_m is

$$N' = \int d^2 r \, |\psi_m|^2 = 2\pi \int_0^\infty d\rho \, \rho \, |\psi_m|^2 = 2\pi \int_0^k d\kappa \, \kappa^{-1} |f(k, \kappa)|^2. \tag{2.35}$$

In the focal plane the Philbin wavefunction is $\frac{\sin k\rho}{k\rho}$, and the integral $\int_0^\infty d\rho\rho \, |\psi_m|^2$ is logarithmically divergent at $\rho \to \infty$. Equivalently, on converting the integral (2.33) to one over κ using $\kappa^2 + q^2 = k^2$ (and thus $\kappa d\kappa + q dq = 0$) gives $f(k, \kappa) = \kappa / kq$, and the last expression in (2.35) is again logarithmically divergent. This difficulty is not removed by the complex shift in z discussed in the previous paragraph. Similar problems arise for electromagnetic beams (to be considered in the next Chapter) constructed using the Philbin wavefunction.

Figure 2.2 shows the Porras, Philbin, and proto beams in the focal plane. The Philbin wavefunction has the smallest zero, at $k\rho = \pi$, followed by the proto-beam at $k\rho \approx 3.83$ (the first zero of $J_1(k\rho)/k\rho$), and then the first Porras zero at $k\rho \approx 5.52$. The Porras beam falls off more rapidly than the other two, by one power of ρ, as explained above.

2.6   PROPERTIES OF $\psi_b(\rho, z)$

We have already noted that beam wavefunctions such as ψ_0 may be generalized to beam families by means of a complex shift along the propagation direction, $z \to z - ib$. As an example of such a beam waveform, which has all of the required physical properties, we shall consider the wavefunction (2.27) (Section 9 of Lekner 2016 [18], Lekner and Andrejic 2018 [20]):

$$\psi_b(\rho, z) = \frac{b^2}{\left[e^{kb}(kb - 1) + 1 \right]} \int_0^k dq \, q \, e^{q(b+iz)} J_0 \left(\rho \sqrt{k^2 - q^2} \right). \tag{2.36}$$

The prefactor in (2.36) normalizes the wavefunction to unity at the origin $\rho = 0$, $z = 0$, for easier comparison with ψ_G given in (2.7), which is also normalized to unity at the origin.

We shall show that, for $e^{kb} \gg 1$ and $\rho^2 \ll b/k$, where the Gaussian waveform (2.7) has some validity, the wavefunction ψ_b corresponds closely to it, provided that also $|z| \ll kb^2$. There

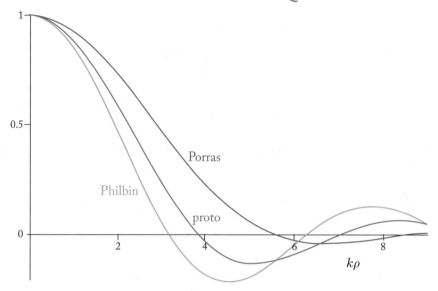

Figure 2.2: The Porras, Philbin and proto beams in the focal plane, drawn in red, green and blue, respectively.

are no constraints on where ψ_b may be used, being an exact solution of the Helmholtz equation. Figure 2.3 shows $\psi_b\,(\rho, z)$ in the focal region around the origin, for kb again set equal to 2, as in Figure 2.1. Note how the different isophase surfaces (except for those on which the phase differs by a multiple of 2π; these can meet anywhere) meet on the zeros of ψ_b, which all lie on circles centered on the beam axis.

On the beam axis $\rho = 0$ we have

$$\psi_b(0, z) = \frac{b^2}{\left[e^{kb}\,(kb - 1) + 1\right]} \int_0^k dq\, q\, e^{q(b+iz)} = \left(\frac{b}{b + iz}\right)^2 \frac{e^{k(b+iz)}\,[k\,(b + iz) - 1] + 1}{\left[e^{kb}\,(kb - 1) + 1\right]}$$

$$(2.37)$$

Equivalently, we may set $z \to z - ib$ in (2.25) and normalize to unity at the origin. An explicit form of ψ_b at a general point (ρ, z) was found in Lekner (2016) [18], using the fact that the expression (2.36) is a cylindrically symmetric non-singular solution of the Helmholtz equation, and may thus be expanded as a sum over products of Legendre polynomials and spherical Bessels,

$$\psi_b\,(\rho, z) = \frac{(kb)^2}{\left[e^{kb}\,(kb - 1) + 1\right]} \sum a_n\, P_n\left(\frac{z - ib}{R}\right) j_n\,(kR)$$

$$R = (z - ib)\sqrt{1 + \rho^2/(z - ib)^2}. \qquad (2.38)$$

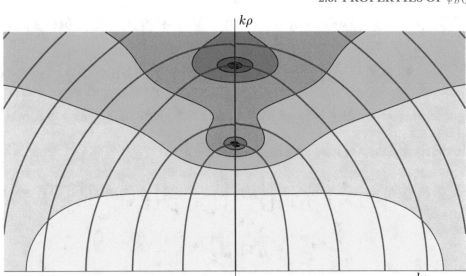

Figure 2.3: $\psi_b(\rho, z)$ in the focal region, plotted for $kb = 2$, for $k|z| \leq 9$, $k\rho \leq 9$. Shading indicates modulus of the wavefunction (logarithmic scale, lighter color indicates larger modulus). The isophase surfaces are shown at intervals of $\pi/2$. The phase is chosen to be zero at the origin. The isophase contours, other than those that are multiples of π, meet on the zeros of $\psi_0(\rho, z)$, two of which are shown, at $k\rho \approx 4.77$ and 7.73.

As in Lekner (2016) [18], R is chosen as a branch of the complex radial coordinate R resulting from an imaginary displacement along the beam axis:

$$r = \sqrt{\rho^2 + z^2} \rightarrow R = \sqrt{\rho^2 + (z - ib)^2}. \tag{2.39}$$

The coefficients a_n in the expansion are given in Appendix B of Lekner (2016) [18], and below. There is only one non-zero odd coefficient, $a_1 = 2i$. The even coefficients we shall rename as $a_{2n} = A_n$, so that

$$\psi_b(\rho, z) = \frac{(kb)^2}{\left[e^{kb}(kb - 1) + 1\right]} \left\{ 2i P_1\left(\frac{z - ib}{R}\right) j_1(kR) + \sum_0^\infty A_n P_{2n}\left(\frac{z - ib}{R}\right) j_{2n}(kR) \right\}. \tag{2.40}$$

The full sequence of the coefficients of the even terms is (note that $(-1)!! = 1$)

$$A_0 = 1, \quad A_n = -\frac{(4n + 1)(2n - 3)!!}{2^n(n + 1)!} = -\frac{2(4n + 1)(2n - 3)!!}{(2n + 2)!!}, \quad n = 1, 2, 3, \ldots \tag{2.41}$$

At the origin $\rho = 0$, $z = 0$ we have $R \to -ib$, $P_n\left(\frac{z-ib}{R}\right) \to P_n(1) = 1$, so (we set $kb = \beta$)

$$\psi_b(0,0) = \frac{\beta^2}{[e^\beta(\beta-1)+1]}\left\{2ij_1(-i\beta) + \sum_0^\infty A_n j_{2n}(-i\beta)\right\}. \tag{2.42}$$

The right-hand side is unity because of the identities given in Equation 9.6 of Lekner (2016) [17].

In the focal plane $z = 0$ we have $R \to -i\sqrt{b^2 - \rho^2}$, so

$$\psi_b(\rho,0) = \frac{\beta^2}{[e^\beta(\beta-1)+1]}\left\{2iP_1\left(\frac{b}{\sqrt{b^2-\rho^2}}\right)j_1\left(-ik\sqrt{b^2-\rho^2}\right)\right.$$
$$\left. + \sum_{n=0}^\infty A_n P_{2n}\left(\frac{b}{\sqrt{b^2-\rho^2}}\right)j_{2n}\left(-ik\sqrt{b^2-\rho^2}\right)\right\}. \tag{2.43}$$

There is a removable singularity on the circle $\rho = b$ in the $z = 0$ plane, since for large X and small x

$$P_n(X) \sim \frac{(2n-1)!!}{n!}X^n, \qquad j_n(x) \sim \frac{x^n}{(2n+1)!!}. \tag{2.44}$$

Thus as $\rho \to b$ the quantity in braces in (2.43) tends to

$$1 + \frac{2\beta}{3} + \sum_{n=1}^\infty A_n \frac{(-\beta^2)^n}{(4n+1)(2n)!} = 1 + \frac{2\beta}{3} - \sum_{n=1}^\infty \frac{(-\beta^2/4)^n}{n!(n+1)!(2n-1)}. \tag{2.45}$$

The final sum may be evaluated as

$$1 - \frac{2}{3}\left[J_0(\beta) + \beta^{-1}J_1(\beta)\right] - \frac{2\beta^2}{3}\left[J_0(\beta) - \beta^{-1}J_1(\beta)\right]$$
$$+ \frac{\pi\beta^2}{3}\left[J_0(\beta)H_1(\beta) - J_1(\beta)H_0(\beta)\right]. \tag{2.46}$$

In this expression J_0, J_1 are the usual Bessel functions, and H_0, H_1 are the related Struve functions (Olver and Maximon (2010) [22], and Chapter 11 of the same book reference).

The focal plane wavefunction (2.43) is real for all ρ. It becomes oscillatory for $\rho > b$: we put $\sqrt{b^2 - \rho^2} = i\sqrt{\rho^2 - b^2}$ to obtain

$$\beta^{-2}\left[e^\beta(\beta-1)+1\right]\psi_b(\rho,0) = \frac{2b}{\sqrt{\rho^2-b^2}}j_1\left(k\sqrt{\rho^2-b^2}\right)$$

$$+ \sum_{n=0}^\infty A_n P_{2n}\left(\frac{-ib}{\sqrt{\rho^2-b^2}}\right)j_{2n}\left(k\sqrt{\rho^2-b^2}\right). \tag{2.47}$$

When $\rho^2 \gg b^2$ the identity (B1) of Lekner 2016 [18] shows that the right-hand side of (2.47) tends to $2J_1(k\rho)/k\rho + 2bj_1(k\rho)/\rho$ to lowest order in b. There is an infinite number of circles on which the wavefunction is zero in the focal plane, where the different isophase surfaces from opposite sides of the focal plane can meet (Section 1.6). In contrast to the Carter beam (2.26), these zeros stay in the focal plane for all kb. Figure 2 of Andrejic and Lekner 2017 [1] shows the variation of the zeros of ψ_b with kb.

The divergence angle of the beam is considered next. We shall see that for large kb it is the same as for the Gaussian beam, $\theta \approx \sqrt{2/kb}$. How the beam expands away from the axis is complicated within the focal region, but simplifies when $|z| \gg b, r \gg b$ and $kr \gg 1$. Let us find the asymptotic form of the un-normalized wavefunction, which from (2.43) is

$$\psi_b(\rho, z) = 2iP_1\left(\frac{z - ib}{R}\right) j_1(kR) + \sum_0^\infty A_n P_{2n}\left(\frac{z - ib}{R}\right) j_{2n}(kR). \tag{2.48}$$

Since $R^2 = r^2 - 2ibz + b^2$, we have $R \approx r - ibz/r$. The argument of the Legendre polynomials is therefore $z/r = \cos\theta$ plus correction terms of order b/z and bz/r^2, which we can neglect. However, $kR \approx kr - ikbz/r$, and the imaginary part is important because it leads to hyperbolic terms in the spherical Bessel functions, for which the asymptotic forms for large $|\zeta|$ are

$$j_1(\zeta) \to -\frac{\cos\zeta}{\zeta}, \qquad j_{2n}(\zeta) \to (-)^n \frac{\sin\zeta}{\zeta}. \tag{2.49}$$

We also know from Equation 3.20 of Lekner (2016) [17] that

$$\sum_0^\infty (-)^n A_n P_{2n}(\cos\theta) = 2|\cos\theta|. \tag{2.50}$$

Hence the asymptotic form of (2A.1) is, on setting $kR \approx kr - ikb\cos\theta$,

$$\frac{2}{kr}\{-i\cos\theta\cos(kr - ikb\cos\theta) + |\cos\theta|\sin(kr - ikb\cos\theta)\}$$
$$= -\frac{2i|\cos\theta|}{kr}\exp\left[ikr\,\mathrm{sgn}(\cos\theta) + kb|\cos\theta|\right]. \tag{2.51}$$

The asymptotic ratio of the moduli (off-axis to on-axis) of ψ_b is therefore

$$|\psi_b(\rho, z)/\psi_b(0, z)| \to \cos^2\theta \, e^{-kb(1-|\cos\theta|)}. \tag{2.52}$$

When kb is large compared to unity the exponent in (2.52) is -1 when $\theta \approx \sqrt{2/kb}$, which is the same as for the Gaussian beam for large kb, given below Equation (2.9). For comparison, for ψ_G the asymptotic ratio of the moduli (off-axis to on-axis) is $e^{-(kb/2)\tan^2\theta}$, quite different from (2.52) except at small θ.

The above analysis is for large values of kb, and does not apply to the proto-beam, for which $kb = 0$. Andrejic (2018) [2] found that the proto-beam has a divergence angle of 45°. This is in accord with the divergence angle of pulses formed by superposition of proto-beams of different frequencies (Lekner 2018) [19].

2.7   FOCAL REGION EXTENT, LOCAL WAVELENGTH

Chapter 8 deals with measures of focal region extent in detail. Here we give an introduction only. We also discuss the concept of local wavelength, which differs (yes, in free space) from the vacuum wavelength wherever the waves differ from plane waves.

There are two lengths (in general) characterizing the extent of the focal region: the transverse localization length L_t, and the longitudinal localization length L_l. There are constraints on how we can define these lengths, because beam wavefunctions decay as a negative power of the distance from the focal region: see for example (2.37) and (2.47). We shall define the longitudinal and transverse extents of the focal region of a scalar beam by

$$L_l = \left\langle |z|^{\frac{1}{2}} \right\rangle^2, \qquad L_t = \left\langle \rho^{\frac{1}{2}} \right\rangle^2 \tag{2.53}$$

$$\left\langle |z|^{\frac{1}{2}} \right\rangle = \frac{\int_0^\infty dz\, z^{\frac{1}{2}}\, |\psi(0,z)|^2}{\int_0^\infty dz\, |\psi(0,z)|^2}, \qquad \left\langle \rho^{\frac{1}{2}} \right\rangle = \frac{\int_0^\infty d\rho\, \rho\, \rho^{\frac{1}{2}}\, |\psi(\rho,0)|^2}{\int_0^\infty d\rho\, \rho\, |\psi(\rho,0)|^2}. \tag{2.54}$$

We begin with the example of the fundamental Gaussian mode, characterized by two parameters $k = \omega/c$ and b:

$$\psi_G = \frac{b}{b+iz} \exp\left\{ ikz - \frac{k\rho^2}{2(b+iz)} \right\}, \qquad |\psi_G|^2 = \frac{b^2}{b^2+z^2} \exp\left\{ -\frac{kb\rho^2}{b^2+z^2} \right\}. \tag{2.55}$$

As we saw in Section 2.2, ψ_G is an *approximate* solution of the Helmholtz equation, valid only when $k^2 \left(b^2 + z^2\right) \gg 1$ and $b^2 + z^2 \gg \rho^2$. It fails in the focal region for small values of kb, and also when $\rho > b$. Note that there are no focal plane zeros of ψ_G. For the Gaussian beam fundamental mode we find

$$L_l = 2b, \qquad L_t = \Gamma^2\left(\frac{5}{4}\right)\left(\frac{b}{k}\right)^{\frac{1}{2}} \approx 0.82 \left(\frac{b}{k}\right)^{\frac{1}{2}}. \tag{2.56}$$

Thus the longitudinal extent of the Gaussian beam is characterized by the length b (called the *diffraction length* or the *Rayleigh length*), while the transverse extent or *beam width* at focus is determined by $(b/k)^{1/2}$.

The proto-beam is the limiting form of two families of exact solutions of the Helmholtz equation; it is the one with the tightest focus within these sets of solutions. Being a limiting form, it has just the one length parameter, k^{-1}. Hence both the longitudinal and transverse extents of the focal region will be proportional to k^{-1}. The axial and focal plane forms are

$$\psi_0(\rho,0) = \frac{2}{k^2} \int_0^k d\kappa\, \kappa J_0(\kappa\rho) = \frac{2J_1(k\rho)}{k\rho} \tag{2.57}$$

$$\psi_0(0,z) = \frac{2}{k^2} \int_0^k dq\, q e^{iqz} = \frac{2}{(kz)^2}\left[(1-ikz)e^{ikz} - 1 \right]. \tag{2.58}$$

We find, on using the axial form (2.57) and the focal plane form (2.58) of ψ_0, that

$$L_l = \frac{288}{25\pi} k^{-1} \approx 3.67\, k^{-1}, \qquad L_t = \frac{4\pi^3}{9\Gamma^8 \left(\frac{3}{4}\right)} k^{-1} \approx 2.71\, k^{-1}. \tag{2.59}$$

Since $k^{-1} = c/\omega = \lambda_0/2\pi$, where λ_0 is the vacuum wavelength at angular frequency ω, we see that the scalar proto-beam is localized in the focal region both longitudinally and transversely to roughly one half of the vacuum wavelength, $L_l \approx 0.58\,\lambda_0$, $L_t \approx 0.43\,\lambda_0$.

Porras (1994) [24] proposed a quality factor, and found its lowest value, which is attained by the Porras beam (2.30). The Porras criterion, a measure of the transverse beam size at its waist, is based on the second moment

$$\langle \rho^2 \rangle = \frac{\int_0^\infty d\rho\, \rho^3\, |\psi(\rho, 0)|^2}{\int_0^\infty d\rho\, \rho\, |\psi(\rho, 0)|^2}. \tag{2.60}$$

In relation to (2.30) it was noted that this measure is convergent only for special choice of wavefunction: in general, exact solutions of the Helmholtz equation will not have a finite second moment. Andrejic (2018) [2] has examined the localization problem in detail and concludes that the proto-beam has the largest peak intensity of all possible physical beam wavefunctions. The Andrejic measures of focal extent are the subject of Chapter 8.

The above discussion of measures of focal region size applies to scalar beams. As we shall see in the next Chapter, many electromagnetic (vector) beams can be formed from the same scalar beam wavefunction. The tightness of focus depends on the polarization properties of the beam, with a radially polarized beam achieving localization to about $0.40\lambda_0$, whereas a linearly polarized beam achieves localization to $0.51\lambda_0$ (Quabis et al. 2000 [25], Dorn et al. 2003 [10]). Here and below $\lambda_0 = 2\pi k^{-1} = 2\pi c\omega^{-1}$; the effective area was determined by the intensity contour at half of maximum. The sharpness of focus of higher-order radially polarized beams is discussed by Kozawa and Sato (2007) [12], and limits of the effective focal volume in multiple-beam light microscopy by Arkhipov and Schulten (2009) [3].

In relation to the localization problem, we note that the *wavelength* within a beam is not λ_0, except far from the focal region. For example, on the beam axis we can track the phase $P(z)$ of the proto-beam: this is the phase of (2.58), namely

$$P(z) = \arctan \frac{\sin kz\ -\ kz\cos kz}{\cos kz\ +\ kz\sin kz\ -\ 1}. \tag{2.61}$$

We define the local wavelength in terms of the variation of phase, extending the concept of wavelength in harmonic plane waves, where it is the distance over which the phase changes by 2π (at a fixed time). We take the large N limit of $P\left(z + \frac{\lambda}{N}\right) - P(z) = \frac{2\pi}{N}$, namely

$$\lambda(z) = \frac{2\pi}{\partial_z P}. \tag{2.62}$$

For the scalar proto-beam, differentiation of (2.61) gives

$$\frac{k\,\lambda(z)}{2\pi} = \frac{(kz)^2 - 2kz\sin kz + 2(1-\cos kz)}{(kz)^2 - kz\sin kz} = \frac{3}{2} - \frac{(kz)^2}{120} + O(kz)^4. \tag{2.63}$$

Thus $\lambda(z) \approx \frac{3}{2}\lambda_0$ near the center of the focal region of the proto-beam. (Far from the focal region $\lambda(z) \to \lambda_0$.) The longer wavelength in the focal region can be seen in Figure 1.1. Its existence is related to the cumulative phase shift of π across the focal region, noted by Gouy in 1890 [11].

For vector beams there is a set of isophase surfaces associated with each of the components of $\boldsymbol{E}$ and $\boldsymbol{B}$. There can be up to six such sets, and each set of isophase surfaces has associated with it a local wavelength given by $\lambda(\boldsymbol{r}) = 2\pi/|\nabla P|$.

2.8   BEAM WAVEFUNCTIONS DERIVED FROM $\psi_0(\rho,z)$

The Helmholtz equation is unchanged by translation of x, y or z. We have used this fact to generate the beam family $\psi_b(\rho,z)$ from $\psi_0(\rho,z)$ in Section 2.6 by means of a complex translation along the beam direction, $z \to z - ib$. The translation invariance also implies that operation with ∂_x, ∂_y or ∂_z will give a solution: these differential operators commute with $\nabla^2 + k^2$. So do the derivative combinations $\partial_x \pm i\partial_y = e^{\pm i\phi}(\partial_\rho \pm i\rho^{-1}\partial_\phi)$, used in Lekner (2008) [16] to construct spinning wave packets from one with zero angular momentum. If Ψ_0 is any $m=0$ solution of the Helmholtz equation, the following are also solutions, with $m = 1$, 2, 3 and 4:

$$\begin{aligned}
\Psi_1 &= e^{i\phi}\partial_\rho\Psi_0 \\
\Psi_2 &= e^{2i\phi}(\partial_\rho^2 - \rho^{-1}\partial_\rho)\Psi_0 \\
\Psi_3 &= e^{3i\phi}(\partial_\rho^3 - 3\rho^{-1}\partial_\rho^2 + 3\rho^{-2}\partial_\rho)\Psi_0 \\
\Psi_4 &= e^{4i\phi}(\partial_\rho^4 - 6\rho^{-1}\partial_\rho^3 + 15\rho^{-2}\partial_\rho^2 - 15\rho^{-3}\partial_\rho)\Psi_0.
\end{aligned} \tag{2.64}$$

We shall first look at an $m = 1$ beam derived from the proto-beam wavefunction ψ_0. The derivative combination $\partial_x + i\partial_y = e^{i\phi}(\partial_\rho + i\rho^{-1}\partial_\phi)$ acting on ψ_0 gives the solution $e^{i\phi}\partial_\rho\psi_0$ of the Helmholtz equation. Also $\partial_\rho J_0(\kappa\rho) = -\kappa J_1(\kappa\rho)$. To keep the wavefunction dimensionless we shall multiply by k^{-1}, and define

$$\psi_1(\rho,\phi,z) = -k^{-1}e^{i\phi}\partial_\rho\psi_0 = e^{i\phi}\int_0^k d\kappa\, f_1(k,\kappa)\, e^{iqz} J_1(\kappa\rho). \tag{2.65}$$

The wavenumber weight function of ψ_1 is thus

$$f_1(k,\kappa) = \kappa k^{-1} f_0(k,\kappa) = 2\kappa^2/k^3. \tag{2.66}$$

Note that $\psi_1(\rho,\phi,z)$ is of the generalized Bessel beam form (2.14), and that it is zero on the beam axis. Further application of the operator $\partial_x + i\partial_y = e^{i\phi}(\partial_\rho + i\rho^{-1}\partial_\phi)$ will produce a beam wavefunction with $m = 2$, and so on. All such beam wavefunctions will be 'hollow,' that is zero on the beam axis $\rho = 0$. In particular, for the $m = 2$ wavefunction we take

$$
\begin{aligned}
\psi_2\left(\rho, \phi, z\right) &= -k^{-1} e^{i\phi}\left(\partial_\rho + i\rho^{-1}\partial_\phi\right)\psi_1 \\
&= -k^{-1} e^{2i\phi}\left(\partial_\rho - \rho^{-1}\right)\int_0^k d\kappa\, f_1\left(k, \kappa\right) e^{iqz} J_1\left(\kappa\rho\right) \\
&= k^{-1} e^{2i\phi}\int_0^k d\kappa\, f_1\left(k, \kappa\right) e^{iqz}\kappa\, J_2\left(\kappa\rho\right).
\end{aligned}
\tag{2.67}
$$

We can write this as

$$
\begin{aligned}
\psi_2\left(\rho, \phi, z\right) &= e^{2i\phi}\int_0^k d\kappa\, f_2\left(k, \kappa\right) e^{iqz} J_2\left(\kappa\rho\right), \\
f_2\left(k, \kappa\right) &= \kappa k^{-1} f_1\left(k, \kappa\right) = 2\kappa^3/k^4.
\end{aligned}
\tag{2.68}
$$

Note that the beam functions ψ_1 and ψ_2 obtained in this way are of the form (2.14), but with different wavenumber weight functions $f(k, \kappa)$.

The remainder of this section explores some of the properties of $\psi_m(\rho, \phi, z)$. First we look at ψ_1 in the focal plane, $z = 0$. We know $\psi_0\left(\rho, 0\right) = 2J_1(k\rho)/k\rho$, so by application of $-k^{-1} e^{i\phi}\partial_\rho\psi_0$, or by evaluation of the integral in (2.65), we have

$$
\psi_1\left(\rho, \phi, 0\right) = 2e^{i\phi}(k\rho)^{-2}\left\{2J_1\left(k\rho\right) - k\rho J_0(k\rho)\right\} = 2e^{i\phi}(k\rho)^{-1} J_2\left(k\rho\right).
\tag{2.69}
$$

There is an infinity of wavefunction zeros in the focal plane, beginning with $k\rho = 0$, and then at $k\rho \approx 5.1356, 8.4172, \ldots$. The amplitude of ψ_1 in the focal plane increases linearly with ρ for $k\rho$ small.

For the $m = 2$ wavefunction we find

$$
\psi_2\left(\rho, \phi, 0\right) = 2e^{2i\phi}(k\rho)^{-1} J_3\left(k\rho\right).
\tag{2.70}
$$

The infinity of zeros in the focal plane (after the zero at the origin) begins with $k\rho \approx 6.380, 9.761, \ldots$. The amplitude of ψ_2 in the focal plane increases quadratically with ρ for small $k\rho$.

For general m the weight functions and focal plane forms are

$$
\begin{aligned}
f_m\left(k, \kappa\right) &= \left(\frac{\kappa}{k}\right)^m \qquad f_0\left(k, \kappa\right) = \left(\frac{\kappa}{k}\right)^m \frac{2\kappa}{k^2}, \\
\psi_m\left(\rho, \phi, 0\right) &= 2e^{im\phi}(k\rho)^{-1} J_{m+1}\left(k\rho\right).
\end{aligned}
\tag{2.71}
$$

Figure 2.4 compares ψ_0, ψ_1 and ψ_2 in the focal plane.

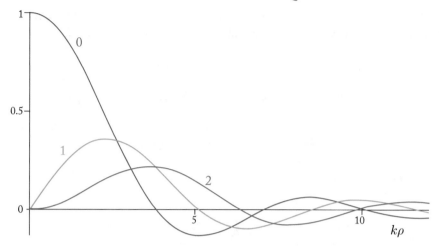

Figure 2.4: The beam functions ψ_0, ψ_1 and ψ_2 in the focal plane (phase factors $e^{im\phi}$ omitted). ψ_0, ψ_1 and ψ_2 are shown in red, green and blue. All of the beams with $m > 0$ are hollow, that is they have zero amplitude on the beam axis. All of the beams have all of their zeros in the focal plane.

2A   APPENDIX: CYLINDRICAL INTEGRALS RELATED TO SPHERICAL SUMS

If we restrict consideration to beam wavefunctions which do not depend on the azimuthal angle ϕ, the cylindrically symmetric solutions of the Helmholtz equation may be expressed either as an integral over the cylindrical Bessel function J_0, or as a sum over a product of spherical Bessel functions j_n and Legendre polynomials P_n:

$$\Psi(r,k) = \int_0^k d\kappa\, f(k,\kappa)\, e^{iz\sqrt{k^2-\kappa^2}}\, J_0(\kappa\rho) = \sum_0^{\infty} a_n(k) j_n(kr) P_n(\cos\theta). \qquad (2A.1)$$

(An example of both expressions for the same beam function is provided by the proto-beam in Appendix 2B, where the coefficients a_n are made explicit. The same coefficients apply to the family ψ_b discussed in Section 2.6.)

To find the coefficients a_n in terms of the wavenumber weight function, we shall use the results ($C = \cos\theta$)

$$\int_{-1}^{1} dC\, P_n(C) P_m(C) = \frac{2\delta_{nm}}{2n+1}, \qquad \int_0^{\infty} dr\, r^2\, j_n(kr) j_n(k'r) = \frac{\pi}{2k^2}\delta(k-k'). \qquad (2A.2)$$

The latter formula follows from the Hankel inversion relation (2.16), since $j_n(\zeta) = \sqrt{\frac{\pi}{2\zeta}} J_{n+\frac{1}{2}}(\zeta)$:

$$\int_0^\infty d\rho\, \rho\, J_\nu(k\rho)\, J_\nu(k'\rho) = k^{-1}\delta(k-k'). \qquad (2A.3)$$

We wish to relate the coefficients $a_n(k)$ of the spherical coordinate expansion to the weight function $f(k,\kappa)$ in the cylindrical coordinate integral. Let us take $\rho = 0$ in (2A.1), that is, equate the on-axis values. On the axis we have $r = |z|$, $P_n(\cos\theta) = [\text{sgn}(z)]^n$, so

$$\int_0^k d\kappa\, f(k,\kappa)\, e^{iz\sqrt{k^2-\kappa^2}} = \sum_0^\infty a_n(k) j_n(k|z|)\left[\text{sgn}(z)\right]^n = \sum_0^\infty a_n(k) j_n(kz). \qquad (2A.4)$$

We change to the variable $q = \sqrt{k^2 - \kappa^2}$, with $\kappa d\kappa + q dq = 0$, and use the auxiliary function $h(k,q) = q\kappa^{-1} f(k,\kappa)$. The left side of (2A.4) becomes, in terms of the dimensionless variables $\eta = q/k$ $(0 \le \eta \le 1)$,

$$\int_0^k dq\, h(k,q)\, e^{iqz} = k\int_0^1 d\eta\, h(k,k\eta)\, e^{i\eta kz}. \qquad (2A.5)$$

We now operate on (2A.4) with $\int_{-\infty}^\infty dz\, j_m(kz)$, $\zeta = kz$, and use the known Fourier transform of spherical Bessels (Olver and Maximon 2010 [22], Equation 10.59.1)

$$\int_{-\infty}^\infty d\zeta\, e^{i\eta\zeta}\, j_n(\zeta) = \begin{cases} \pi i^n P_n(\eta), & -1 < \eta < 1 \\ \dfrac{\pi}{2}(\pm i)^n, & \eta = \pm 1 \\ 0, & \pm\eta > 1. \end{cases} \qquad (2A.6)$$

On the left-hand side $\int_{-\infty}^\infty dz\, j_m(kz)$ times (2A.5) gives $\pi i^n \int_0^1 d\eta\, P_n(\eta)\, h(k,k\eta)$.
On the right-hand side of $\int_{-\infty}^\infty dz\, j_m(kz)$ operating on (2A.4) we use the orthogonality condition

$$\int_{-\infty}^\infty d\zeta\, j_m(\zeta)\, j_n(\zeta) = \frac{\pi\delta_{nm}}{2n+1}. \qquad (2A.7)$$

The transformed (2A.4) thus evaluates the Bessel–Legendre series coefficients:

$$a_n(k) = (2n+1)\, i^n k \int_0^1 d\eta\, P_n(\eta)\, h(k,k\eta). \qquad (2A.8)$$

For the proto-beam $h(k,q) = \frac{2q}{k^2} = \frac{2\eta}{k}$, and we find the coefficients $a_0 = 1$, $a_1 = 2i$, $a_2 = -\frac{5}{4},\ldots$, with all odd coefficients apart from a_1 being zero. The general formula for the even coefficients is given in Appendix 2B.

For the Philbin beam $h(k,q) = k^{-1}$, and the reverse occurs: $a_0 = 1$ is the only non-zero even coefficient, and $a_1 = \frac{3i}{2}$, $a_3 = \frac{7i}{8}$, $a_5 = \frac{11i}{16},\ldots$. The general formula for the odd coefficients is

$$a_{2m+1} = \frac{i\,(4m+3)\,(2m-1)!!}{(2m+2)!!} \qquad (m = 0,\, 1,\, 2,\ldots) \qquad (2A.9)$$

2B   APPENDIX: PROPERTIES OF THE PROTO-BEAM WAVEFUNCTION

The proto-beam is defined in (2.28)

$$\psi_0\left(\rho, z\right) = \frac{2}{k^2} \int_0^k d\kappa \,\kappa J_0\left(\kappa\rho\right) e^{iqz} = \frac{2}{k^2} \int_0^k dq \, q J_0\left(\kappa\rho\right) e^{iqz} \quad (\kappa^2 + q^2 = k^2). \quad (2\text{B}.1)$$

The prefactor $2/k^2$ normalizes the beam wavefunction to unity at the origin: $\psi_0\left(0,0\right) = 1$. In the focal plane $z = 0$ we have

$$\psi_0\left(\rho, 0\right) = \frac{2}{k^2} \int_0^k d\kappa \,\kappa J_0\left(\kappa\rho\right) = \frac{2 J_1(k\rho)}{k\rho} \quad (2\text{B}.2)$$

[the integration follows from $z^n J_{n-1}\left(z\right) = \frac{d}{dz}\left[z^n J_n\left(z\right)\right], \; n = 1, 2, \ldots$]. We note that the Airy formula for the amplitude of a wave focused by a circular lens of aperture radius a and focal length f is of the form (2B.2), but with $k\rho$ replaced by $k\rho\left(a/f\right)$ (Born and Wolf 1965 [7], Section 8.8). The beam wavefunction has an infinite number of circles of zeros in the focal plane, the necessity for which has been argued on topological grounds in Section 1.6.

The form of the wavefunction on the beam axis $\rho = 0$ is also simple:

$$\psi_0\left(0, z\right) = \frac{2}{k^2} \int_0^k dq \, q e^{iqz} = \frac{2}{(kz)^2}\left[(1-ikz)\, e^{ikz} - 1\right]$$

$$= \frac{2}{(kz)^2}\left[kz\sin kz + \cos kz - 1\right] + \frac{2i}{(kz)^2}\left[\sin kz - kz\cos kz\right]. \quad (2\text{B}.3)$$

To obtain one form of the complete beam wavefunction, we use the Bessel integral (Watson, 1944 [30], Section 2.21)

$$J_0\left(\rho\sqrt{k^2 - q^2}\right) = \frac{1}{2\pi} \int_0^{2\pi} d\phi \; e^{q\rho\cos\phi + ik\rho\sin\phi} \quad . \quad (2\text{B}.4)$$

Substitution into (2B.1) gives

$$\psi_0\left(\rho, z\right) = \frac{1}{\pi} \int_0^{2\pi} d\phi \, e^{-ik\rho\sin\phi} \frac{1}{k^2} \int_0^k dq \, q \, e^{iq(z + i\rho\cos\phi)}$$

$$= \frac{1}{\pi} \int_0^{2\pi} d\phi \, e^{-ik\rho\sin\phi} \frac{1 - (1-\eta)\, e^{\eta}}{\eta^2}, \qquad \eta = ik\left(z + i\rho\cos\phi\right). \quad (2\text{B}.5)$$

In Appendix A of Lekner (2016) [18] it is shown from (2B.5) that the proto-beam is given by

$$\psi_0(\rho, z) = F(u, v) + i G(u, v) \qquad u = k\left(r - z\right), \qquad v = k\rho. \quad (2\text{B}.6)$$

The functions F, G are expressed in terms of the Bessel functions $J_0 = J_0(v)$, $J_1 = J_1(v)$, and the Lommel functions of two variables (Watson 1944 [30], Section 16.5) $U_0 = U_0(u, v)$, $U_1 = U_1(u, v)$. It is convenient to define the auxiliary variable $w = \frac{u^2 + v^2}{2u}$, and also $C = \cos w$, $S = \sin w$. Then

$$F(u, v) = \frac{8u^2 v J_1}{(u^2 + v^2)^2} + \frac{4u(u^2 - v^2)}{(u^2 + v^2)^3} \left[2u\,(2U_0 - J_0 - C) + (u^2 + v^2)\,(2U_1 - S) \right] \quad (2B.7)$$

$$G(u, v) = \frac{4u(u^2 - v^2)}{(u^2 + v^2)^3} \left[(u^2 + v^2)\,C - 2uS \right]. \quad (2B.8)$$

In the focal plane $z = 0$, $u = k(r - z) \to k\rho = v$, so $G \to 0$ and $F \to 2J_1(v)/v$, in accord with (2B.2). From the defining Equation 2B.1 we see that

$$\psi_0(\rho, z)^* = \psi_0(\rho, -z), \quad \text{so} \quad F(\rho, -z) = F(\rho, z), \quad \text{and} \quad G(\rho, -z) = -G(\rho, z). \quad (2B.9)$$

It follows from (2B.9) that the surfaces of constant phase are antisymmetric about the focal plane:

$$P(\rho, z) = \arctan \frac{G(\rho, z)}{F(\rho, z)} = -P(\rho, -z). \quad (2B.10)$$

It is useful to express the real and imaginary parts F, G in terms of the radial coordinate $r = \sqrt{\rho^2 + z^2}$ and the axial coordinate z. Since $u = k\,(r - z)$, $v = k\rho$,

$$u^2 + v^2 = (2kr)\,u, \qquad u^2 - v^2 = -(2kz)\,u, \qquad w = kr. \quad (2B.11)$$

The imaginary part of the beam wavefunction is then seen to be odd in z, in accord with (2B.9):

$$G(\rho, z) = \frac{2kz}{(kr)^3}[\sin kr - kr \cos kr] = \frac{2z}{r} j_1(kr) = 2j_1(kr)P_1(\cos \theta). \quad (2B.12)$$

In Appendix 2A we showed that cylindrically symmetric solutions of the Helmholtz equation may be expressed as a sum over a product of spherical Bessel functions j_n and Legendre polynomials P_n. For the real part F of ψ_0, which we know to be even, the sum contains only the even Legendre polynomials:

$$F(r, \theta) = \sum_0^\infty A_n P_{2n}(\cos \theta)\, j_{2n}(kr). \quad (2B.13)$$

The coefficients A_n can be evaluated using (2A.8) above. Here we use the fact that

$$F\left(r, \frac{\pi}{2}\right) = \sum_0^\infty A_n P_{2n}(0)\, j_{2n}(kr) = \frac{2J_1(kr)}{kr}. \quad (2B.14)$$

It is known that $P_{2n}(0) = (-)^n (2n-1)!! / (2n)!!$ (Lebedev 1972 [13], Equation 4.2.7). Let us first solve for $B_n = A_n P_{2n}(0)$. The expansion in powers of kr of both sides of (2B.14) are known,

$$\frac{2J_1(x)}{x} = \sum_{n=0}^{\infty} (-)^n \frac{\left(\frac{x}{2}\right)^{2n}}{n!\,(n+1)!}, \qquad j_n(x) = x^n \sum_{p=0}^{\infty} \frac{\left(-\frac{x^2}{2}\right)^p}{p!\,(2n+2p+1)!!}. \tag{2B.15}$$

Equating the coefficients of $(kr)^{2n}$ in the two expressions on the right-hand side of (2B.14) gives

$$\frac{(-)^n}{2^{2n}n!\,(n+1)!} = \sum_{m+p=n} \frac{B_m(-)^p}{2^p p!\,(4m+2p+1)!!} = \sum_{m=0}^{n} \frac{(-)^{n-m} B_m}{2^{n-m}(n-m)!(2n+2m+1)!!}. \tag{2B.16}$$

Hence

$$\frac{1}{2^n n!\,(n+1)!} = \sum_{m=0}^{n} \frac{2^m (-)^m B_m}{(n-m)!(2n+2m+1)!!} = \sum_{m=0}^{n} \frac{(2m-1)!!\,A_m}{(n-m)!m!(2n+2m+1)!!}. \tag{2B.17}$$

Viewed as a matrix equation, the vector A is multiplied by a matrix in echelon form: we can solve the $n=0$ equation to find $A_0 = 1$, then solve the $n=1$ equation (which contains A_0, A_1) for A_1, and so on. We find

$$A_0 = 1, \quad A_1 = -\frac{5}{4}, \quad A_2 = -\frac{3}{8}, \quad A_3 = -\frac{13}{64}, \quad A_4 = -\frac{17}{128}, \quad A_5 = -\frac{49}{512}, \dots \tag{2B.18}$$

The full sequence is (note that $(-1)!! = 1$)

$$A_0 = 1, \qquad A_n = -\frac{(4n+1)(2n-3)!!}{2^n (n+1)!} = -\frac{2(4n+1)(2n-3)!!}{(2n+2)!!}, \qquad n = 1,\ 2,\ 3, \dots. \tag{2B.19}$$

The result (2B.19) follows from the identities (for $n = 1,\ 2,\ 3, \dots$)

$$\frac{1}{n!}\left\{ \frac{1}{(2n+1)!!} - \frac{2}{(2n+2)!!} \right\} = \sum_{m=1}^{n} \frac{(4m+1)(2m-1)!!(2m-3)!!}{2^m (m+1)!m!(n-m)!(2n+2m+1)!!}. \tag{2B.20}$$

The asymptotic form of $F(r,\theta)$ for large kr is found from $j_{2n}(kr) \to (-)^n \sin kr / kr$:

$$F(r,\theta) \to \frac{\sin kr}{kr} \sum_0^{\infty} (-)^n A_n P_{2n}(\cos\theta) = \frac{\sin kr}{kr} 2\,|\cos\theta|. \tag{2B.21}$$

The sum follows from the fact that $|\cos\theta|$ is even about $\theta = \pi/2$, and can therefore be expanded in the orthogonal set of even order Legendre polynomials. The relation

$\sum_0^\infty (-)^n A_n P_{2n} (\cos \theta) = 2 |\cos \theta|$ can be verified from

$$\int_0^1 dC \, P_{2n}(C) \, P_{2m}(C) = \frac{\delta_{nm}}{4n+1}, \qquad \int_0^1 dC \, P_{2n}(C) \, C = \frac{(-)^{n+1} (2n-3)!!}{(2n+2)!!} \qquad (n \geq 1).$$
$$(2B.22)$$

To sum up the results so far: the imaginary part of ψ_0 is known in closed form, as shown in (2B.12). The real part is however known as an integral, in terms of Bessel functions and Lommel functions of two variables, or as an infinite sum over products of spherical Bessel functions j_n and Legendre polynomials P_n. Since the Lommel functions of two variables are themselves infinite sums over Bessel functions, the compact form (2B.7) is not useful numerically. We therefore seek a more tractable form of

$$F(\rho, z) = \frac{2}{k^2} \int_0^k d\kappa \, \kappa J_0 (\kappa \rho) \cos qz. \qquad (2B.23)$$

We expand the cosine in powers of $qz = z\sqrt{k^2 - \kappa^2}$ and integrate term by term, using Sonine's first finite integral (Watson 1944 [30], Section 12.11), which transcribes to

$$\int_0^k d\kappa \, \kappa \left(k^2 - \kappa^2\right)^n J_0 (\kappa \rho) = 2^n n! \, k^{2n+2} \frac{J_{n+1}(k\rho)}{(k\rho)^{n+1}}. \qquad (2B.24)$$

This gives us

$$F(\rho, z) = 2 \sum_0^\infty \frac{(-)^n 2^n n! (kz)^{2n}}{(2n)!} \frac{J_{n+1}(k\rho)}{(k\rho)^{n+1}}$$
$$= \frac{2}{k\rho} \sum_0^\infty \frac{(-)^n 2^n n!}{(2n)!} \left(\frac{kz^2}{\rho}\right)^n J_{n+1}(k\rho) = \frac{2}{k\rho} \sum_0^\infty \frac{(-)^n}{(2n-1)!!} \left(\frac{kz^2}{\rho}\right)^n J_{n+1}(k\rho).$$
$$(2B.25)$$

We compare (2B.25) with Lommel's expansion of a Bessel function containing a square root, specifically Equation (16) of Watson (1944) [30], Section 5.22, in which we write $\lambda^2 = 1 + 2z^2/\rho^2$:

$$\frac{2J_1(k\sqrt{\rho^2 + 2z^2})}{k\sqrt{\rho^2 + 2z^2}} = \frac{2}{k\rho} \sum_0^\infty \frac{(-)^n}{n!} \left(\frac{kz^2}{\rho}\right)^n J_{n+1}(k\rho). \qquad (2B.26)$$

The series (2B.25) and (2B.26) are similar: in fact the first two terms ($n = 0, 1$) are the same. We have thus shown that

$$F(\rho, z) = \frac{2J_1(k\sqrt{\rho^2 + 2z^2})}{k\sqrt{\rho^2 + 2z^2}} + \frac{2}{k\rho} \sum_2^\infty (-)^n \left\{\frac{1}{(2n-1)!!} - \frac{1}{n!}\right\} \left(\frac{kz^2}{\rho}\right)^n J_{n+1}(k\rho). \quad (2B.27)$$

The alternating series converges rapidly for small $|z|$, since the lead term is of fourth degree in z.

Laplace transform method: we use the Bessel integral (2B.4) to write

$$F(\rho, z) = \frac{1}{2\pi} \int_0^k dq \, q \, \cos qz \int_0^{2\pi} d\phi \, e^{q\rho \cos\phi + ik\rho \sin\phi} \,. \tag{2B.28}$$

The Laplace transform of $\frac{1}{2}k^2 F(\rho, z)$ is obtained by operating with $\int_0^\infty dk e^{-ks}$. The result is

$$\int_0^\infty dk e^{-ks} \frac{1}{2} k^2 F(\rho, z) = \frac{1}{R} \frac{R^2 - z^2}{(R^2 + z^2)^2}, \qquad R^2 = s^2 + \rho^2. \tag{2B.29}$$

The inverse Laplace transform gives us

$$\frac{1}{2} k^2 F(\rho, z) = \frac{\rho^2}{r^3} \int_0^k d\kappa \, J_0(\kappa\rho) \sin(k - \kappa) r$$

$$+ \frac{z^2}{r^2} \int_0^k d\kappa \, J_0(\kappa\rho) (k - \kappa) \cos(k - \kappa) r. \tag{2B.30}$$

In the focal plane ($z = 0$, $r = \rho$) we regain (2B.2):

$$F(\rho, 0) = \frac{2}{k^2\rho} \int_0^k d\kappa \, J_0(\kappa\rho) \sin(k - \kappa) \rho = \frac{2J_1(k\rho)}{k\rho}. \tag{2B.31}$$

On the beam axis ($\rho = 0$, $r = |z|$) we have, as in (2B.3),

$$F(0, z) = \frac{2}{k^2} \int_0^k d\kappa \, (k - \kappa) \cos(k - \kappa) z = \frac{2}{(kz)^2} [kz \, \sin kz + \cos kz - 1]. \tag{2B.32}$$

For general ρ, z the function F is expressed in terms of the integrals ($n = 0, 1$)

$$S_n = \int_0^k d\kappa \, \kappa^n \, J_0(\kappa\rho) \sin \kappa r \,, \qquad C_n = \int_0^k d\kappa \, \kappa^n \, J_0(\kappa\rho) \cos \kappa r \,. \tag{2B.33}$$

Of these we require closed forms only for $C_0(\rho, z)$, $S_0(\rho, z)$ since

$$C_1(\rho, z) = \frac{r}{z} \partial_z S_0, \qquad S_1(\rho, z) = -\frac{r}{z} \partial_z C_0. \tag{2B.34}$$

In terms of the S_n, C_n, the real part of the proto-beam wavefunction is given by

$$\frac{1}{2} k^2 F(\rho, z) = \frac{\rho^2}{r^3} \{C_0 \sin kr - S_0 \cos kr\}$$

$$+ \frac{z^2}{r^2} \{[kC_0 - C_1] \cos kr + [kS_0 - S_1] \sin kr\} \tag{2B.35}$$

$$= r^{-1} \sin^2\theta \, \{C_0 \sin kr - S_0 \cos kr\}$$

$$+ \cos^2\theta \, \{[kC_0 - C_1] \cos kr + [kS_0 - S_1] \sin kr\}.$$

In the *focal plane* the derivatives of ψ_0 can be obtained by differentiation of (2B.1), setting $z = 0$, and then direct integration. We shall list all up to second order. The odd-order derivatives with respect to z are expressed in terms of the spherical Bessel functions j_n, and are imaginary.

$$\psi_0 = 2(k\rho)^{-1} J_1(k\rho)$$

$$
\begin{aligned}
k^{-1}\partial_\rho \psi &= 2(k\rho)^{-2}\left[k\rho J_0(k\rho) - 2J_1(k\rho)\right] = -2(k\rho)^{-1} J_2(k\rho) \\
k^{-2}\partial_\rho^2 \psi &= 2(k\rho)^{-3}\left[(6 - (k\rho)^2)J_1(k\rho) - 3k\rho J_0(k\rho)\right] \\
k^{-1}\partial_z \psi &= 2i(k\rho)^{-3}\left[\sin k\rho - k\rho \cos k\rho\right] = 2i(k\rho)^{-1} j_1(k\rho) \\
k^{-2}\partial_\rho \partial_z \psi &= 2i(k\rho)^{-4}\left[\left((k\rho)^2 - 3\right)\sin k\rho + 3k\rho \cos k\rho\right] = -2i(k\rho)^{-1} j_2(k\rho) \\
k^{-2}\partial_z^2 \psi &= 4(k\rho)^{-3}\left[k\rho J_0(k\rho) - 2J_1(k\rho)\right] = -4(k\rho)^{-2} J_2(k\rho).
\end{aligned}
\tag{2B.36}
$$

2.11   REFERENCES

[1] Andrejic, P. and Lekner, J. 2017. Topology of phase and polarization singularities in focal regions, *Journal of Optics*, 19(105609):8. DOI: 10.1088/2040-8986/aa895d. 17, 23

[2] Andrejic, P. 2018. Convergent measure of focal extent, and largest peak intensity for non-paraxial beams, *Journal of Optics*, 20(075610):10. DOI: 10.1088/2040-8986/aaca6b. 23, 25

[3] Arkhipov, A. and Schulten, K. 2009. Limits for reduction of effective focal volume in multiple-beam light microscopy, *Optics Express*, 17:2861–2870. DOI: 10.1364/oe.17.002861. 25

[4] Baranov, A. S. 2006. On series containing products of Legendre polynomials, *Mathematical Notes*, 80:167–174. DOI: 10.1007/s11006-006-0124-5.

[5] Bateman, H. 1904. The solution of partial differential equations by means of definite integrals, *Proc. of the London Mathematical Society*, 1:451–458. DOI: 10.1112/plms/s2-1.1.451. 15

[6] Berry, M. V. 1998. Wave dislocation reactions in non-paraxial Gaussian beams, *Journal of Modern Optics*, 45:1845–1858. DOI: 10.1080/09500349808231706. 17

[7] Born, M. and Wolf, E. 1965. *Principles of Optics*, 3rd ed., Pergamon, Oxford. 30

[8] Carter, W. H. 1973. Anomalies in the field of a Gaussian beam near focus, *Optics Communications*, 64:491–495. DOI: 10.1016/0030-4018(73)90012-6. 17

[9] Deschamps, G. A. 1971. Gaussian beam as bundle of complex rays, *Electronics Letters*, 7:684–685. DOI: 10.1049/el:19710467. 11

[10] Dorn, R., Quabis, S., and Leuchs, G. 2003. Sharper focus for a radially polarized light beam, *Physical Review Letters*, 91(233901):1–4. DOI: 10.1103/physrevlett.91.233901. 25

[11] Gouy, L. G. 1890. Sur une propriété nouvelle des ondes lumineuses, *Comptes Rendus de Academie des Sciences Paris*, 110:1251–1253. 12, 26

[12] Kozawa, Y. and Sato, S. 2007. Sharper focal spot formed by higher-order radially polarized laser beams, *Journal of the Optical Society of America*, 24:1793–1798. DOI: 10.1364/josaa.24.001793. 25

[13] Lebedev, N. N. 1972. *Special Functions and their Applications*, Dover, New York. 32

[14] Lekner, J. 2001. TM, TE, and "TEM" beam modes: Exact solutions and their problems, *Journal of Optics A: Pure and Applied Optics*, 3:407–412. DOI: 10.1088/1464-4258/3/5/314. 12, 13

[15] Lekner, J. 2004. Invariants of three types of generalized Bessel beams, *Journal of Optics A: Pure and Applied Optics*, 6:837–843. DOI: 10.1088/1464-4258/6/9/004.

[16] Lekner, J. 2008. Rotating wavepackets, *European Journal of Physics*, 29:1121–1125. DOI: 10.1088/0143-0807/29/5/025. 26

[17] Lekner, J. 2016. Tight focusing of light beams: A set of exact solutions, *Proc. of the Royal Society A*, 472(20160538):17. DOI: 10.1098/rspa.2016.0538. 17, 22, 23

[18] Lekner, J. 2016. *Theory of Reflection*, 2nd ed., Springer. DOI: 10.1007/978-3-319-23627-8. 16, 19, 20, 21, 23, 30

[19] Lekner, J. 2018. Electromagnetic pulses, localized and causal, *Proc. of the Royal Society A*, 474(20170655):17. DOI: 10.1098/rspa.2017.0655. 23

[20] Lekner, J. and Andrejic, P. 2018. Nonexistence of exact solutions agreeing with the Gaussian beam on the beam axis or in the focal plane, *Optics Communications*, 407:22–26. DOI: 10.1016/j.optcom.2017.08.071. 14, 19

[21] Nye, J. F. 1998. Unfolding of higher-order wave dislocations, *Journal of the Optical Society of America*, 15:1132–1138. DOI: 10.1364/josaa.15.001132. 17

[22] Olver, F. W. J. and Maximon, L. C. 2010. *Bessel Functions*, Chapter 10 of *NIST Handbook of Mathematical Functions*, Olver, F. W. J. et al., Eds., Cambridge University Press. 22, 29

[23] Philbin, T. G. 2018. Some exact solutions for light beams, *Journal of Optics*, 20(105603):15. DOI: 10.1088/2040-8986/aade6d. 18

[24] Porras, M. A. 1994. The best quality optical beam beyond the paraxial approximation, *Optics Communications*, 111:338–349. DOI: 10.1016/0030-4018(94)90475-8. 18, 25

[25] Quabis, S., Dorn, R., Eberler, M., Glöckl, O., and Leuchs, G. 2000. Focusing light to a tighter spot, *Optical Communications*, 179:1–7. DOI: 10.1016/s0030-4018(99)00729-4. 25

[26] Sheppard, C. J. R. and Shagati, S. 1998. Beam modes beyond the paraxial approximation: a scalar treatment. *Phys. Rev.*, A57:2971–2979. 11

[27] Todhunter, I. 1911. *Spherical Trigonometry*, Macmillan, London.

[28] Ulanowski, Z. and Ludlow, I. K. 2000. Scalar field of nonparaxial Gaussian beams, *Opt. Lett.*, 25:1792–1794. 11

[29] Vinti, J. P. 1951. Note on a series of products of three Legendre polynomials, *Proc. of the American Mathematical Society*, 2:19–23. DOI: 10.1090/s0002-9939-1951-0043266-1.

[30] Watson, G. N. 1944. *Theory of Bessel Functions*, Cambridge University Press. 15, 16, 30, 31, 33

[31] Zangwill, A. 2013. *Modern Electrodynamics*, Cambridge University Press. DOI: 10.1017/cbo9781139034777. 12

CHAPTER 3

Electromagnetic Beams

3.1 E AND B FROM SOLUTIONS OF THE WAVE EQUATION

We summarize here the basic equations of electromagnetism from Section 1.2, as applied to monochromatic beams. The Maxwell's equations are (in Gaussian units) in free-space and with $\partial_{ct} = c^{-1}\partial/\partial t$,

$$\nabla \cdot \boldsymbol{B} = 0 \qquad\qquad \nabla \cdot \boldsymbol{E} = 0$$
$$\nabla \times \boldsymbol{E} + \partial_{ct}\boldsymbol{B} = 0 \qquad\qquad \nabla \times \boldsymbol{B} - \partial_{ct}\boldsymbol{E} = 0. \tag{3.1}$$

Electric and magnetic fields can be expressed in terms of the vector potential $\boldsymbol{A}(\boldsymbol{r},t)$ and scalar potential $V(\boldsymbol{r},t)$ via

$$\boldsymbol{E} = -\nabla V - \partial_{ct}\boldsymbol{A}, \qquad \boldsymbol{B} = \nabla \times \boldsymbol{A}. \tag{3.2}$$

With these substitutions the source-free Maxwell equations $\nabla \cdot \boldsymbol{B} = 0$, $\nabla \times \boldsymbol{E} + \partial_{ct}\boldsymbol{B} = 0$ are satisfied automatically. If further $\boldsymbol{A}$ and V satisfy the Lorenz condition $\nabla \cdot \boldsymbol{A} + \partial_{ct}V = 0$, substitution of (3.2) into Maxwell's free space equations (of which the curl equations couple $\boldsymbol{E}$ and $\boldsymbol{B}$), decouples the vector and the scalar potentials:

$$\nabla^2\boldsymbol{A} - \partial_{ct}^2\boldsymbol{A} = 0, \qquad \nabla^2 V - \partial_{ct}^2 V = 0. \tag{3.3}$$

For monochromatic beams of angular frequency $\omega = ck$ the time dependence is in the factor $e^{-i\omega t}$ in the case of complex field amplitudes, and either $\cos\omega t$ or $\sin\omega t$ for real fields. In all cases the wave equations in (3.3) become vector or scalar Helmholtz equations of the form

$$\nabla^2\psi + k^2\psi = 0. \tag{3.4}$$

When we are dealing with *complex field amplitudes*, and the corresponding complex potentials, the Lorenz condition $\nabla \cdot \boldsymbol{A} + \partial_{ct}V = 0$ becomes

$$\nabla \cdot \boldsymbol{A} - ikV = 0, \qquad \text{so} \qquad V = (ik)^{-1}\nabla \cdot \boldsymbol{A}. \tag{3.5}$$

Hence, for monochromatic complex fields and with the Lorenz condition satisfied, both the electric and the magnetic complex field amplitudes fields are given in terms of the complex vector potential $\boldsymbol{A}(\boldsymbol{r})$:

$$\boldsymbol{E}(\boldsymbol{r}) = \frac{i}{k}[\nabla(\nabla \cdot \boldsymbol{A}) + k^2\boldsymbol{A}], \qquad \boldsymbol{B}(\boldsymbol{r}) = \nabla \times \boldsymbol{A}. \tag{3.6}$$

It is understood, in the use of complex amplitudes, that the real fields are obtained by taking the real or imaginary parts of the corresponding amplitude times $e^{-i\omega t}$. For example,

$$
\begin{aligned}
\boldsymbol{E}(\boldsymbol{r},t) &= Re\left\{\boldsymbol{E}(\boldsymbol{r})e^{-i\omega t}\right\} = Re\left\{[\boldsymbol{E}_r(\boldsymbol{r}) + i\,\boldsymbol{E}_i(\boldsymbol{r})]e^{-i\omega t}\right\} \\
&= \boldsymbol{E}_r(\boldsymbol{r})\cos\omega t + \boldsymbol{E}_i(\boldsymbol{r})\sin\omega t.
\end{aligned}
\tag{3.7}
$$

The energy, momentum and angular momentum densities of an electromagnetic field in free space are, with real fields $\boldsymbol{E}(\boldsymbol{r},t)$, $\boldsymbol{B}(\boldsymbol{r},t)$, given by

$$
u(\boldsymbol{r},t) = \frac{1}{8\pi}\left(E^2 + B^2\right), \qquad \boldsymbol{p}(\boldsymbol{r},t) = \frac{1}{4\pi c}\boldsymbol{E}\times\boldsymbol{B}, \qquad \boldsymbol{j}(\boldsymbol{r},t) = \boldsymbol{r}\times\boldsymbol{p}(\boldsymbol{r},t).
\tag{3.8}
$$

The energy flux density is $\boldsymbol{S} = c^2\boldsymbol{p}$. In free space the Maxwell equations (3.1) imply the energy conservation relation

$$
\nabla\cdot\boldsymbol{S} + \partial_t u = 0, \qquad \text{or} \qquad c\nabla\cdot\boldsymbol{p} + \partial_{ct}u = 0.
\tag{3.9}
$$

The average of $u(\boldsymbol{r},t)$ over one period $2\pi/\omega$ is, in terms of the complex field amplitudes,

$$
\bar{u}(\boldsymbol{r}) = \frac{1}{16\pi}\left\{\boldsymbol{E}(\boldsymbol{r})\cdot\boldsymbol{E}^*(\boldsymbol{r}) + \boldsymbol{B}(\boldsymbol{r})\cdot\boldsymbol{B}^*(\boldsymbol{r})\right\} = \frac{1}{16\pi}\left\{E_r^2 + E_i^2 + B_r^2 + B_i^2\right\}.
\tag{3.10}
$$

Likewise the cycle-averaged momentum density is

$$
\bar{\boldsymbol{p}}(\boldsymbol{r}) = \frac{1}{16\pi c}\left[\boldsymbol{E}(\boldsymbol{r})\times\boldsymbol{B}^*(\boldsymbol{r}) + \boldsymbol{E}^*(\boldsymbol{r})\times\boldsymbol{B}(\boldsymbol{r})\right] = \frac{1}{8\pi c}\left[\boldsymbol{E}_r\times\boldsymbol{B}_r + \boldsymbol{E}_i\times\boldsymbol{B}_i\right].
\tag{3.11}
$$

The real and imaginary parts of the complex field amplitudes are different functions, for example the electric field may be given by (3.7) or by $Im\left\{\boldsymbol{E}(\boldsymbol{r})e^{-i\omega t}\right\} = \boldsymbol{E}_i(\boldsymbol{r})\cos\omega t - \boldsymbol{E}_r(\boldsymbol{r})\sin\omega t$. The beams derived from the real and the imaginary parts differ in their space-time dependence: they are in phase quadrature. However, as is clear from the symmetry between real and imaginary parts in (3.10) and (3.11), the cycle-averaged energy and momentum densities are the same for the two beams, at each point in space.

3.2   TM AND TE BEAMS

The simplest beam has a vector potential with only the longitudinal component non-zero (Davis and Patsakos 1981 [1]). This gives us the TM (transverse magnetic) beam, as we shall see shortly.

Let $\psi(\boldsymbol{r})$ be a wavefunction satisfying the Helmholtz equation, and A_0 a real constant. Then

$$
\boldsymbol{A}_{TM} = A_0\,[0,0,\psi] = A_0(0,0,\psi).
\tag{3.12}
$$

We shall give expressions in both Cartesian $[x,y,z]$ and cylindrical polar (ρ,ϕ,z) coordinates. It is assumed that the ϕ dependence of ψ is in the factor $e^{im\phi}$, as in (2.14), repeated below:

$$
\psi(\boldsymbol{r}) = e^{im\phi}\int_0^k d\kappa\, f(k,\kappa)\,e^{iqz}\,J_m(\kappa\rho), \qquad q = \sqrt{k^2 - \kappa^2}.
\tag{3.13}
$$

In that case we have

$$\partial_x = \cos\phi\ \partial_\rho - \rho^{-1}\sin\phi\ \partial_\phi \to \cos\phi\ \partial_\rho - im\rho^{-1}\sin\phi$$
$$\partial_y = \sin\phi\ \partial_\rho + \rho^{-1}\cos\phi\ \partial_\phi \to \sin\phi\ \partial_\rho + im\rho^{-1}\cos\phi \quad (3.14)$$
$$\partial_x + i\,\partial_y = e^{i\phi}\left(\partial_\rho + i\rho^{-1}\partial_\phi\right) \to e^{i\phi}\left(\partial_\rho - m\rho^{-1}\right).$$

The complex magnetic amplitude has no longitudinal component (hence the name TM, transverse magnetic):

$$\boldsymbol{B}\,(\boldsymbol{r}) = \nabla \times \boldsymbol{A} = A_0\left[\partial_y, -\partial_x,\ 0\right]\psi = A_0\left(im\rho^{-1}, -\partial_\rho,\ 0\right)\psi. \quad (3.15)$$

The complex electric amplitude is

$$\boldsymbol{E}\,(\boldsymbol{r}) = \frac{i}{k}\left[\nabla\,(\nabla\cdot\boldsymbol{A}) + k^2\boldsymbol{A}\right] = ik^{-1}A_0\left[\partial_x\partial_z,\ \partial_y\partial_z,\ \partial_z^2 + k^2\right]\psi.$$
$$= ik^{-1}A_0\left(\partial_\rho\partial_z,\ im\rho^{-1}\partial_z,\ \partial_z^2 + k^2\right)\psi. \quad (3.16)$$

The cycle-averaged energy and momentum densities $\bar{u}\,(\boldsymbol{r})$, $\bar{\boldsymbol{p}}\,(\boldsymbol{r})$ may now be found from (3.10) and (3.11). The energy density is

$$\bar{u} = \frac{A_0^2}{16\pi k^2}\left\{\left|\partial_\rho\partial_z\psi\right|^2 + \left|\partial_z^2\psi + k^2\psi\right|^2 + k^2\left|\partial_\rho\psi\right|^2 + m^2\rho^{-2}\left[\left|\partial_z\psi\right|^2 + k^2|\psi|^2\right]\right\}. \quad (3.17)$$

Of the momentum, we are interested in the longitudinal and azimuthal components. The beam has net momentum along its direction of propagation (the z direction in this book), with zero net momentum transversely. The longitudinal angular momentum density is

$$j_z = (\boldsymbol{r} \times \boldsymbol{p}) = xp_y - yp_x = \rho p_\phi. \quad (3.18)$$

The component of interest is along z: it is intrinsic to the beam, unchanged by a shift of origin, since the net transverse components of momentum are zero. To make this explicit, consider the shift $\boldsymbol{r} \to \boldsymbol{r} - \boldsymbol{a}$. The new angular momentum is $(\boldsymbol{r} - \boldsymbol{a}) \times \boldsymbol{p} = \boldsymbol{r} \times \boldsymbol{p} - \boldsymbol{a} \times \boldsymbol{p}$. The z component of $\boldsymbol{a} \times \boldsymbol{p}$ is $a_x p_y - a_y p_x$, with null transverse integral at fixed z : $\int d^2 r p_x = 0 = \int d^2 r p_y$. From (3.11) and the fields (3.15) and (3.16) we find the longitudinal and azimuthal momentum components to be

$$c\overline{p_z} = \frac{A_0^2}{8\pi k}Im\left\{\partial_\rho\psi^*\partial_\rho\partial_z\psi + m^2\rho^{-2}\psi^*\partial_z\psi\right\} \quad (3.19)$$

$$c\overline{p_\phi} = \frac{A_0^2}{8\pi k}m\rho^{-1}\left[Re\left\{\psi^*\partial_z^2\psi\right\} + k^2|\psi|^2\right]. \quad (3.20)$$

The Maxwell equations are unchanged by the *duality transformation* $\boldsymbol{E} \to \boldsymbol{B}$, $\boldsymbol{B} \to -\boldsymbol{E}$. Thus we can produce a TE (transverse electric) beam satisfying the Maxwell equations by applying the duality transformation to (3.15) and (3.16). The duality transformation leaves the energy and momentum densities unchanged, so the energy, momentum and angular momentum of TE and TM beams based on the same wavefunction are identical. However, the polarization properties differ: for example, if $m = 0$ the TM beam *magnetic* field is linearly polarized, since from (3.15) we see that it has only one (azimuthal) component. Likewise, the TE beam with $m = 0$ has its *electric* field azimuthal, and linearly polarized (Lekner 2004 [5]). Polarization is discussed in detail in Chapter 4.

3.3   INVARIANT QUANTITIES OF TM AND TE BEAMS

The conservation of energy Equation (3.9), $c\nabla \cdot \boldsymbol{p} + \partial_{ct} u = 0$, has the cycle-average

$$\nabla \cdot \bar{\boldsymbol{p}} = \partial_x \bar{p}_x + \partial_y \bar{p}_y + \partial_z \bar{p}_z = 0. \tag{3.21}$$

Applying $\int d^2 r = \int_{-\infty}^{\infty} dx \int_{-\infty}^{\infty} dy = \int_0^{\infty} d\rho \rho \int_0^{2\pi} d\phi$ to (3.20) gives, for transversely finite beams propagating in the z direction (Lekner 2004a [5])

$$\partial_z \int d^2 r \, \bar{p}_z = 0, \qquad \text{or} \qquad P_z' = \int d^2 r \, \bar{p}_z = \text{constant}. \tag{3.22}$$

The reason for the notation P_z' is that $dP_z' = P_z' dz$ is the total z-component momentum contained in a transverse slice of the beam, of thickness dz. Equation (3.22) states that the momentum content per unit length, along the direction of net propagation of the beam, is constant along the length of the beam (in the absence of scattering and absorption).

Let us now calculate P_z', for TM and TE beams based on the general wavefunction (3.13). From (3.19) we need to integrate over the imaginary part of $\partial_\rho \psi^* \partial_\rho \partial_z \psi + m^2 \rho^{-2} \psi^* \partial_z \psi$. For ψ of the form (3.13) this expression is independent of ϕ (as are all the terms in $\hat{u}$ and $\overline{p_\phi}$). We shall write

$$\psi^* = e^{-im\phi} \int_0^k d\kappa \, f^*(k, \kappa) \, e^{-iqz} J_m(\kappa\rho),$$

$$\psi = e^{im\phi} \int_0^k d\kappa' f(k, \kappa') \, e^{iq'z} J_m(\kappa'\rho). \tag{3.23}$$

Differentiation of ψ with respect to z brings down a factor of iq'. In the differentiation with respect to ρ we use (Watson 1944 [6], Section 2.12)

$$2J_m'(\zeta) = J_{m-1}(\zeta) - J_{m+1}(\zeta), \qquad \frac{2m}{\zeta} J_m(\zeta) = J_{m-1}(\zeta) + J_{m+1}(\zeta). \tag{3.24}$$

From these recurrence relations it follows that

$$\partial_\rho J_m(\kappa\rho) = m\rho^{-1} J_m(\kappa\rho) - \kappa J_{m+1}(\kappa\rho). \tag{3.25}$$

After the elimination of derivatives by use of (3.25), the momentum density $\bar{p}_z$ contains products of Bessel functions and inverse powers of ρ, in the combination $T_0 - T_1 + 2T_2$, where

$$T_0 = \kappa\kappa' J_{m+1}(\kappa\rho) J_{m+1}(\kappa'\rho),$$
$$T_1 = m\rho^{-1}\left\{\kappa' J_{m+1}(\kappa\rho) J_m(\kappa'\rho) + \kappa J_m(\kappa\rho) J_{m+1}(\kappa'\rho)\right\}, \qquad (3.26)$$
$$T_2 = m^2\rho^{-2} J_m(\kappa\rho) J_m(\kappa'\rho).$$

The integration over ρ of T_0 can be performed using Hankel's inversion formula (Watson 1944 [6], Section 14.4; see also Lekner 2004 [5], Appendix A), which may be written as

$$\int_0^\infty d\rho\,\rho\,J_m(\kappa\rho)\,J_m(\kappa'\rho) = \kappa^{-1}\delta(\kappa - \kappa') \qquad (\kappa, \kappa' > 0). \qquad (3.27)$$

The integration over ρ of products of Bessel functions of the same order thus selects $\kappa' = \kappa$. The integration over ϕ gives 2π. The T_0 contribution to cP'_z, which turns out to be total value as we shall see shortly, is therefore

$$cP'_z = \frac{A_0^2}{4k}\int_0^k d\kappa\,\kappa\,q\,|f(k,\kappa)|^2. \qquad (3.28)$$

The T_2 term can be evaluated by a special case of the Weber–Schafheitlin integral (Watson 1944 [6], Section 13.42 (1)), namely

$$\int_0^\infty d\rho\,\rho^{-1}\,J_m(\kappa\rho)\,J_m(\kappa'\rho) = (2m)^{-1}\left[\frac{\min(\kappa, \kappa')}{\max(\kappa, \kappa')}\right]^m. \qquad (3.29)$$

The integral over the term T_1 can be evaluated in terms of the more general Weber–Schafheitlin integral (Watson 1944 [6], Section 13.4 (2)), resulting in an exact cancellation in the term $2T_2 - T_1$.

The integration over the energy density is similar but a little more complicated. The details are in Appendix A of Lekner (2004) [5]. The energy per unit length of the beam reduces to

$$U' = \frac{A_0^2}{4}\int_0^k d\kappa\,\kappa\,|f(k,\kappa)|^2. \qquad (3.30)$$

In the calculation of the z component of the angular momentum content per unit length of the beam we need to integrate over $j_z = \rho\overline{p_\phi}$, where the azimuthal momentum density is given in (3.20). From (3.23) we have

$$\psi^*\partial_z^2\psi + k^2|\psi|^2$$
$$= \int_0^k d\kappa\,f^*(k,\kappa)e^{-iqz}J_m(\kappa\rho)\int_0^k d\kappa'\,f(k,\kappa')(\kappa')^2 e^{iq'z}J_m(\kappa'\rho). \qquad (3.31)$$

The integral of J_z' can be evaluated by the Hankel inversion formula (3.27), because the extra factor of ρ in j_z is cancelled by the ρ^{-1} (3.23). The result is

$$cJ_z' = \frac{A_0^2}{4k} m \int_0^k d\kappa \, \kappa \, |f(k,\kappa)|^2. \tag{3.32}$$

Thus the total energy, momentum and angular momentum per unit length of TM and TE beams have been expressed as integrals over the wavenumber weight function $f(k,\kappa)$. (Lekner 2004 [5] calculates further tensor invariants of TM and TE beams arising from conservation of momentum and angular momentum.) The expressions obtained above are summarized in

$$\begin{bmatrix} U' \\ cP_z' \\ cJ_z' \end{bmatrix} = \frac{A_0^2}{4k} \int_0^k d\kappa \, \kappa \, |f(k,\kappa)|^2 \begin{bmatrix} k \\ q \\ m \end{bmatrix} \qquad \text{(TM or TE).} \tag{3.33}$$

These results are based entirely on classical electrodynamics, but show that an electromagnetic TE or TM beam can be viewed as a superposition of photons with energies $\hbar ck$, z component of momentum $\hbar k_z = \hbar q$, and z component of angular momentum $\hbar m$. In either the classical or the quantum picture, $mU' = kcJ_z' = \omega J_z'$.

We shall calculate the energy, momentum and angular momentum per unit length for the TM or TE beams based on ψ_0 defined in (2.28), which has $f(k,\kappa) = 2k^{-2}\kappa$. The formulae (3.33) give

$$U' = \frac{A_0^2}{4}, \qquad cP_z' = \frac{A_0^2}{4}\frac{8}{15}, \qquad J_z' = 0 \qquad \text{(TM or TE } m = 0). \tag{3.34}$$

The value of cP_z' is smaller than that of U', in accord with the inequality of Section 1.4.

The values in (3.34) can be checked by direct integration in the focal plane. In the $z = 0$ plane all the derivatives of ψ_0 may be evaluated analytically, directly from the defining integral (2.28). We shall list all up to second order. The odd-order derivatives with respect to z are expressed in terms of the spherical Bessel functions j_n, and are imaginary.

$$\psi_0 = 2(k\rho)^{-1}J_1(k\rho)$$

$$k^{-1}\partial_\rho\psi = 2(k\rho)^{-2}[k\rho J_0(k\rho) - 2J_1(k\rho)] = -2(k\rho)^{-1}J_2(k\rho)$$

$$k^{-2}\partial_\rho^2\psi = 2(k\rho)^{-3}\left[(6 - (k\rho)^2)J_1(k\rho) - 3k\rho J_0(k\rho)\right]$$

$$k^{-1}\partial_z\psi = 2i(k\rho)^{-3}[\sin k\rho - k\rho \cos k\rho] = 2i(k\rho)^{-1}j_1(k\rho) \tag{3.35}$$

$$k^{-2}\partial_\rho\partial_z\psi = 2i(k\rho)^{-4}\left[\left((k\rho)^2 - 3\right)\sin k\rho + 3k\rho \cos k\rho\right] = -2i(k\rho)^{-1}j_2(k\rho)$$

$$k^{-2}\partial_z^2\psi = 4(k\rho)^{-3}[k\rho J_0(k\rho) - 2J_1(k\rho)] = -4(k\rho)^{-2}J_2(k\rho).$$

The integrals needed in the evaluation of the energy, momentum and angular momentum per unit length of the beam are all special cases of the critical form of the Weber–Schafheitlin integral (Watson 1944 [6], Section 13.41),

$$\int_0^\infty dX \frac{J_\mu(X) J_\nu(X)}{X^\lambda} = \frac{\Gamma(\lambda)\Gamma\left(\dfrac{\mu + \nu - \lambda + 1}{2}\right)}{2^\lambda \Gamma\left(\dfrac{\lambda + \mu + \nu + 1}{2}\right)\Gamma\left(\dfrac{\lambda - \mu + \nu + 1}{2}\right)\Gamma\left(\dfrac{\lambda + \mu - \nu + 1}{2}\right)},$$

$$Re\,(\mu + \nu + 1) > Re\,(\lambda) > 0.$$

$$(3.36)$$

With the use of $j_n(z) = \sqrt{\pi/2z}\,J_{n+\frac{1}{2}}(z)$, $z\Gamma(z) = \Gamma(z+1)$, $\Gamma(1) = 1$, $\Gamma\left(\frac{1}{2}\right) = \sqrt{\pi}$ we find, for example,

$$\int_0^\infty dX\, X^{-1} J_1(X)^2 = \frac{1}{2}, \quad \int_0^\infty dX\, X^{-1} J_2(X)^2 = \frac{1}{4},$$

$$\int_0^\infty dX\, X^{-2} J_1(X) J_2(x) = \frac{1}{8}, \quad \int_0^\infty dX\, X^{-3} J_2(X)^2 = \frac{1}{24},$$

$$\int_0^\infty dX\, X^{-3} J_1(X)\,[2J_1(X) - XJ_0(X)] = \frac{1}{8}, \quad \int_0^\infty dX\, j_1(X) J_2(X) = \frac{2}{3},$$

$$\int_0^\infty dX\, X^{-1} j_1(X) J_1(X) = \frac{1}{3}, \quad \int_0^\infty dX\, X^{-1} j_0(X) J_2(X) = \frac{1}{6}, \quad (3.37)$$

$$\int_0^\infty dX\, j_0(X) j_1(X) = \frac{1}{2}, \quad \int_0^\infty dX\, X^{-1} j_1(X)^2 = \frac{1}{4},$$

$$\int_0^\infty dX\, X^{-1}\, j_2(X)^2 = \frac{1}{12} \quad \int_0^\infty dX\, X^{-1} j_2(X) J_2(X) = \frac{2}{15},$$

$$\int_0^\infty dX\, X^{-2} j_1(X) J_2(X) = \frac{1}{10}, \quad \int_0^\infty dX\, X^{-2} j_2(X) J_1(X) = \frac{1}{15}.$$

The Weber–Schafheitlin integral enables evaluation of the energy, momentum and angular momentum per unit length of the beam. The results agree with those in (3.34) which were found by integrating over the wavenumber weight function.

Next we consider the TM or TE beams based on the wavefunction ψ_1 of Section 2.8, namely

$$\psi_1 = -k^{-1} e^{i\phi} \partial_\rho \psi_0 = e^{i\phi} \int_0^k d\kappa\, f_1(k, \kappa)\, e^{iqz} J_1(\kappa\rho). \qquad (3.38)$$

The weight function of ψ_1 is

$$f_1(k, \kappa) = \kappa k^{-1} f_0(k, \kappa) = 2\kappa^2/k^3. \qquad (3.39)$$

The formulae (3.33) now give

$$U' = \frac{A_0^2}{6}, \qquad cP_z' = \frac{A_0^2}{6}\frac{16}{35}, \qquad ckJ_z' = \frac{A_0^2}{6} \qquad \text{(TM or TE, } m=1\text{)}. \qquad (3.40)$$

The ratio cP_z'/U', which was $8/15 \approx 0.533$ for the TM and TE beams based on ψ_0, is now slightly smaller, $16/35 \approx 0.457$. The sign of J_z' would have reversed if we had used $\psi_{-1} = -k^{-1}e^{-i\phi}\partial_\rho\psi_0$ instead of ψ_1.

For the TM and TE beams the ratio $ckJ_z'/U' = m$, as noted below (3.33). For the TM and TE beams based on the functions ψ_m derived from ψ_0 (Section 2.8) the ratio cP_z'/E' decreases with m: for $m = 0,\ 1,\ 2$ it is $8/15$, $16/35$, $128/315$ respectively. The expression for general m is

$$\frac{cP_z'}{E'} = \frac{2^{m+2}\,(m+2)!}{(2m+5)!!} = \frac{1}{2}\sqrt{\frac{\pi}{m}} + O\left(m^{-\frac{3}{2}}\right). \qquad (3.41)$$

These results follow from the weight functions or from the focal plane forms given in (2.71) and repeated here:

$$f_m(k,\kappa) = \left(\frac{-\kappa}{k}\right)^m f_0(k,\kappa) = \frac{2\kappa}{k^2}\left(\frac{-\kappa}{k}\right)^m, \quad \psi_m(\rho,\phi,0) = 2e^{im\phi}(k\rho)^{-1}J_{m+1}(k\rho). \qquad (3.42)$$

3.4   TM AND TE BEAMS BASED ON ψ_0

In the previous Section we have already discussed invariant quantities of the TM and TE beams based on the functions ψ_m derived from ψ_0. In this and the following Sections we show explicitly the electric and magnetic fields of the beams based on ψ_0 and on ψ_1, repectively, as well as the associated energy and momentum densities.

For the $m = 0$ TM beam we see from (3.15) that the magnetic field is purely azimuthal. The magnetic field lines are circles centered on the beam axis. In the focal plane $z = 0$ the magnetic and electric complex amplitudes are functions of ρ alone. We find from (3.35) that

$$B_\phi(\rho,0) = -\partial_\rho\psi_0 = -2E_0(k\rho)^{-2}[k\rho J_0(k\rho) - 2J_1(k\rho)] = 2E_0(k\rho)^{-1}J_2(k\rho). \qquad (3.43)$$

We have set $kA_0 = E_0$, a field amplitude. The electric field complex amplitude has radial and longitudinal components, from (3.16). In the focal plane these are, again from (3.35),

$$E_\rho(\rho,0) = 2E_0(k\rho)^{-1}j_2(k\rho)$$

$$E_z(\rho,0) = 2iE_0\left\{2(k\rho)^{-3}[k\rho J_0(k\rho) - 2J_1(k\rho)] + (k\rho)^{-1}J_1(k\rho)\right\}. \qquad (3.44)$$

We take the real parts of these amplitudes times $e^{-i\omega t}$; the real fields thus have the time dependence $\cos\omega t$ for $B_\phi(\rho,0)$, $E_\rho(\rho,0)$ and $\sin\omega t$ for $E_z(\rho,0)$. Figure 3.1 shows how the

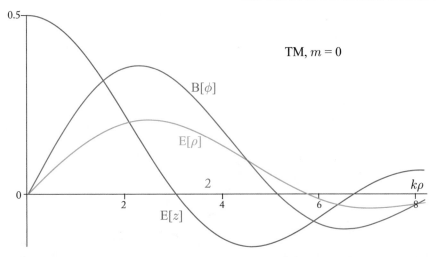

Figure 3.1: **E** and **B** of a TM beam based on ψ_0, in the focal plane. $B_\phi\,(\rho, 0)\,,\;E_\rho\,(\rho, 0)$ oscillate in phase, and the longitudinal component $E_z\,(\rho, 0)$ is in phase quadrature to the transverse components. For the TE beam the fields are obtained by the duality transformation $\boldsymbol{E} \to \boldsymbol{B}, \boldsymbol{B} \to -\boldsymbol{E}$. The ordinate gives the field amplitudes divided by $E_0 = kA_0$.

non-zero components of **E** and **B** vary in focal plane. Since B_ϕ is the only non-zero magnetic field component, the momentum density and the energy flux density (both proportional to $\boldsymbol{E} \times \boldsymbol{B}$) will be zero when $B_\phi = 0$. In the focal plane this happens at $k\rho = 0,\; 5.1356\ldots.$

Figure 3.2 shows the cycle-averaged energy and momentum densities in the focal plane. The densities are the same for TM and TE beams; the fields are duals of each other. The cycle-averaged energy and momentum densities are also the same whether the real or the imaginary parts of the complex amplitudes are used in their computation. For example we could take $\boldsymbol{E}\,(\boldsymbol{r}, t) = Im\left\{\boldsymbol{E}(\boldsymbol{r})e^{-i\omega t}\right\} = Im\left\{[\boldsymbol{E}_r(\boldsymbol{r}) + i\,\boldsymbol{E}_i(\boldsymbol{r})]e^{-i\omega t}\right\} = \boldsymbol{E}_i(\boldsymbol{r})\,\cos\omega t - \boldsymbol{E}_r(\boldsymbol{r})\,\sin\omega t$ instead of (3.7) and get the same cycle-averaged densities. We see that $\bar{u} \geq c\overline{p_z}$, in accord with the inequality of Section 1.4. Both the TM and TE beams are hollow in momentum, but not hollow in energy. For example, in the TM beam E_z is not zero when $\rho = 0$.

Figure 3.3 shows a longitudinal section of the cycle-averaged energy density contours and momentum density vector field of the same TM and TE beams. As noted above, the beams are hollow in momentum, but not hollow in energy. Likewise, the zero of B_ϕ in focal plane at $k\rho \approx 5.1356$ gives a zero in momentum, not in energy.

3.5   TM AND TE BEAMS BASED ON ψ_1

The TM and TE beams with $m = 0$ carry no angular momentum: the angular momentum density component of interest is $j_z = \rho p_\phi$ and, for the TM beams, $\boldsymbol{E} \times \boldsymbol{B} = (-E_z B_\phi,\; 0,\; E_\rho B_\phi)$

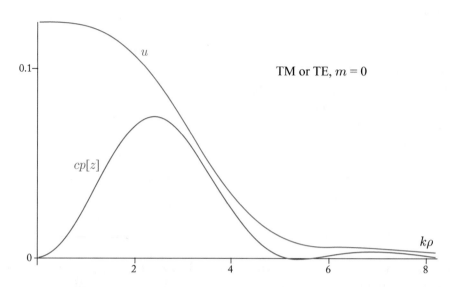

Figure 3.2: The focal plane variation of the cycle-averaged energy and longitudinal momentum density of a TM beam based on ψ_0. Note the zeros in $c\overline{p_z}$, due to zeros in B_ϕ and E_ρ (compare Figure 3.1). The ordinate times $E_0^2/8\pi$ $(E_0 = kA_0)$ gives $\overline{u}, c\overline{p_z}$ and $ck\overline{j_z}$.

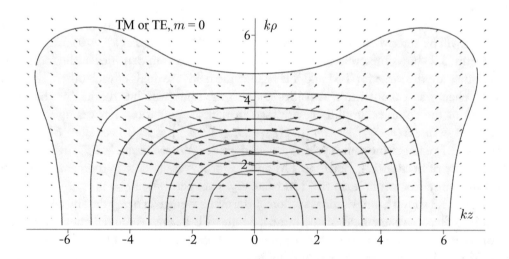

Figure 3.3: Longitudinal section of the cycle-averaged energy density (contours), and the cycle-averaged momentum density vector field (arrows) for the TM and TE beams based on the proto-beam wavefunction ψ_0. The focal region is centered on the origin. A three-dimensional picture is obtained by rotating the figure about the horizontal axis.

when m is zero. For general m the complex field amplitudes are, from (3.15) and (3.16)

$$\boldsymbol{B}(\boldsymbol{r}) = \left(im\rho^{-1}, -\partial_\rho, 0\right)\psi, \qquad \boldsymbol{E}(\boldsymbol{r}) = ik^{-1}A_0\left(\partial_\rho\partial_z, im\rho^{-1}\partial_z, \partial_z^2 + k^2\right)\psi. \qquad (3.45)$$

Now $(\boldsymbol{E}\times\boldsymbol{B})_\phi = E_z B_\rho$ is non-zero, and proportional to m. We wish to explore the properties of TM and TE beams based on the first of the series of hollow beams derived from ψ_0 which was considered in Section 2.8. The wavefunction is

$$\psi_1 = -k^{-1}e^{i\phi}\partial_\rho\psi_0 = 2k^{-3}e^{i\phi}\int_0^k d\kappa\,\kappa^2\,e^{iqz}J_1(\kappa\rho) \qquad \left(\kappa^2 + q^2 = k^2\right). \qquad (3.46)$$

In the focal plane $\psi_1(\rho, \phi, 0) = 2e^{i\phi}(k\rho)^{-1}J_2(k\rho)$, with zeros on the axis $\rho = 0$ and at $k\rho \approx$ 5.1356, 8.4172, ... The complex field amplitudes are (we omit the common factor $kA_0 = E_0$)

$$B_\rho = 2i\,e^{i\phi}(k\rho)^{-2}J_2(k\rho), \quad B_\phi = 2e^{i\phi}(k\rho)^{-2}\left[3J_2(k\rho) - k\rho J_1(k\rho)\right], \quad B_z = 0 \qquad (3.47)$$

$$E_\rho = 2e^{i\phi}(k\rho)^{-5}\left\{\left[(k\rho)^3 - 12k\rho\right]\cos(k\rho) - \left[(k\rho)^2 - 12\right]\sin(k\rho)\right\}$$
$$E_\phi = -2i\,e^{i\phi}(k\rho)^{-2}j_2(k\rho) \qquad\qquad\qquad\qquad\qquad\qquad (3.48)$$
$$E_z = -2i\,e^{i\phi}(k\rho)^{-4}\left\{\left[(k\rho)^3 - 8k\rho\right]J_0(k\rho) - 4\left[(k\rho)^2 - 4\right]J_1(k\rho)\right\}.$$

All the complex amplitudes carry the phase factor $e^{i\phi}$, and are otherwise real or imaginary (in the focal plane). Hence the real fields will have time and azimuthal dependence intertwined in either $\cos(\omega t - \phi)$ or $\sin(\omega t - \phi)$. Apart from these sinusoidal factors the fields have no azimuthal dependence. Figure 3.4 shows the field amplitudes in the focal plane (sinusoidal factors omitted).

From Equations (3.47) the real magnetic field components are $B_\rho = B_1\sin(\omega t - \phi)$, $B_\phi = B_2\cos(\omega t - \phi)$, where $B_1 = 2(k\rho)^{-2}J_2(k\rho)$, $B_2 = 2(k\rho)^{-2}[3J_2(k\rho) - k\rho J_1(k\rho)]$. Hence

$$\left(\frac{B_\rho}{B_1}\right)^2 + \left(\frac{B_\phi}{B_2}\right)^2 = 1. \qquad (3.49)$$

The magnetic field vector $\boldsymbol{B}(\rho, \phi, 0)$ lies in the focal plane, and at any fixed point $(\rho, \phi, 0)$, its endpoint moves in time on an ellipse with semiaxes B_1, B_2 (it is elliptically polarized in general). When either of B_1, B_2 is zero, the magnetic field has linear polarization: the vector $\boldsymbol{B}$ moves on a line as time increases. When $B_1^2 = B_2^2$ the magnetic field is circularly polarized, because the end-point of the vector $\boldsymbol{B}$ moves on a circle as time changes. Polarization will be discussed in detail in Chapter 4.

Next we look at the cycle-averaged energy, momentum and angular momentum densities. These are shown in Figure 3.5; all may be found in closed form from the focal plane field values given in (3.47) and (3.48). There is no azimuthal dependence in the time-averaged densities. The radial momentum component p_ρ is zero in the focal plane. Note that $\bar{u} \geq c\overline{p_z}$, in accord with the inequality of Section 1.4.

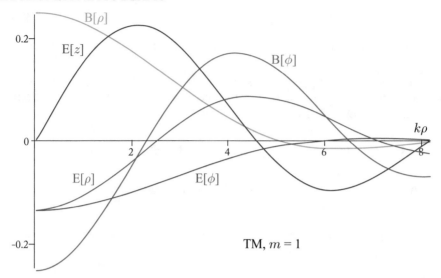

Figure 3.4: **E** and **B** of a TM beam based on ψ_1, in the focal plane. The ordinate gives the field amplitudes divided by $E_0 = kA_0$. All focal plane field components vary with azimuthal angle and time, as $\cos(\omega t - \phi)$ in the case of B_ϕ, E_ρ or $\sin(\omega t - \phi)$ in the case of B_ρ, E_ϕ, E_z. The amplitudes of these sinusoidal factors are plotted.

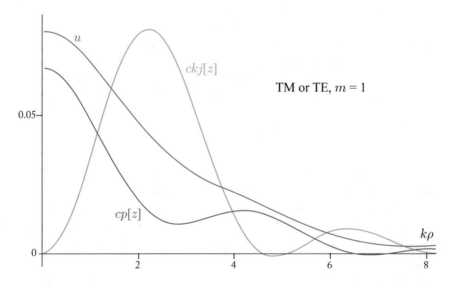

Figure 3.5: The focal plane variation of the cycle-averaged energy, longitudinal momentum density and angular momentum densities of a TM beam based on ψ_1. The ordinate times $E_0^2/8\pi$ gives $\bar{u}$, $c\overline{p_z}$ and $ck\overline{j_z}$. The $m = 1$ TM and TE beams are no longer hollow in momentum, in contrast to those based on ψ_0 (compare Figure 3.2).

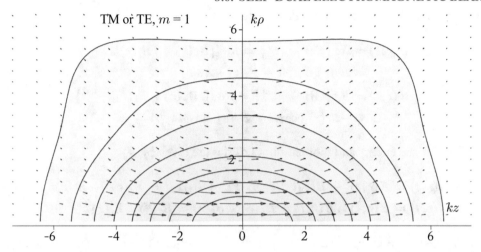

Figure 3.6: Longitudinal section of the cycle-averaged energy density (contours), and the cycle-averaged momentum density vector field (arrows) for the TM and TE beams based on the wavefunction ψ_1. The focal region is centered on the origin. A three-dimensional picture is obtained by rotating the figure about the horizontal axis.

Finally, we look at the energy, momentum and angular momentum in a longitudinal section through the focal region, in Figure 3.6. The wavefunction (3.46) can be written as

$$\psi_1 = 2k^{-3}e^{i\phi} \int_0^k d\kappa \, \kappa^2 \, (\cos qz \; + i \sin qz \,) \, J_1(\kappa\rho) \equiv 2k^{-3}e^{i\phi}(C + iS). \qquad (3.50)$$

C and S stand for the integrals over κ of $\kappa^2 J_1(\kappa\rho)$, times $\cos qz$ and $\sin qz$, respectively. The resulting cycle-averaged energy and momentum densities (3.10) and (3.11) are thus independent of ϕ.

3.6   SELF-DUAL ELECTROMAGNETIC BEAMS

Maxwell's free-space equations are unchanged by the *duality transformation* $\boldsymbol{E} \to \boldsymbol{B}$, $\boldsymbol{B} \to -\boldsymbol{E}$. The duality transformation also leaves the energy and momentum densities unchanged. Any solution of Maxwell's equations has its dual: for example, a TM beam based on ψ_m has the TE beam based on ψ_m as its dual, and vice-versa. These beams are different, but have the same energy and momentum densities, as noted.

Self-dual electromagnetic fields are those unchanged by the duality transformation. They have particularly simple properties because of the symmetry between the electric and magnetic fields. Complex field amplitudes which satisfy $\boldsymbol{E} = \pm i \, \boldsymbol{B}$ give real fields which are self-dual. We have

$$\boldsymbol{E}_r + i \, \boldsymbol{E}_i = \pm i \, (\boldsymbol{B}_r + i \, \boldsymbol{B}_i), \qquad \boldsymbol{E}_r = \mp \boldsymbol{B}_i, \qquad \boldsymbol{E}_i = \pm \boldsymbol{B}_r. \qquad (3.51)$$

For monochromatic complex amplitudes satisfying $E = iB$ the real fields are

$$
\begin{aligned}
B(r,t) &= Re\left\{B(r)e^{-i\omega t}\right\} = Re\left\{[B_r(r) + iB_i(r)]e^{-i\omega t}\right\} \\
&= B_r(r)\cos\omega t + B_i(r)\sin\omega t \\
E(r,t) &= Re\left\{iB(r)e^{-i\omega t}\right\} = Re\left\{[iB_r(r) - B_i(r)]e^{-i\omega t}\right\} \\
&= -B_i(r)\cos\omega t + B_r(r)\sin\omega t
\end{aligned}
\tag{3.52}
$$

When $E = iB$ the energy and momentum densities are given by

$$
\begin{aligned}
8\pi u(r,t) &= E(r,t)^2 + B(r,t)^2 = B_r(r)^2 + B_i(r)^2 = E_r(r)^2 + E_i(r)^2 \\
4\pi c\, p(r,t) &= E(r,t) \times B(r,t) = B_r(r) \times B_i(r) = E_r(r) \times E_i(r).
\end{aligned}
\tag{3.53}
$$

(When $E = -iB$ the energy density is unchanged; the momentum density expression changes sign.) Hence in self-dual monochromatic (or 'steady') beams the *electromagnetic energy and momentum densities do not oscillate in time* (Lekner 2002 [3], Section 4), whereas as we saw in Section 3.1 they normally oscillate at twice the angular frequency of the beam.

We note also that in monochromatic self-dual beams both the electric and magnetic fields are *eigenvectors of curl* in free space (Lekner 2002 [3]). When $E = \pm iB$

$$
\begin{aligned}
0 &= \nabla \times E(r,t) + \partial_{ct} B(r,t) \\
&= \nabla \times (E_r(r)\cos\omega t + E_i(r)\sin\omega t) + \partial_{ct}[B_r(r)\cos\omega t + B_i(r)\sin\omega t] \\
&= \nabla \times [E_r(r)\cos\omega t + E_i(r)\sin\omega t] + k[-B_r(r)\sin\omega t + B_i(r)\cos\omega t] \\
&= \nabla \times [E_r(r)\cos\omega t + E_i(r)\sin\omega t] + k[\mp E_i(r)\sin\omega t \mp E_r(r)\cos\omega t].
\end{aligned}
\tag{3.54}
$$

On equating the coefficients of $\cos\omega t$ and of $\sin\omega t$ we have for the real fields (with $k = \omega/c$)

$$
\nabla \times E_r(r) = \pm k E_r(r), \qquad \nabla \times E_i(r) = \pm k E_i(r).
\tag{3.55}
$$

The same equations follow for the real and imaginary parts of the complex magnetic amplitude. Hence real and imaginary parts of the electric complex amplitude are eigenvectors of curl, and likewise the real and imaginary parts of the magnetic complex amplitude are eigenvectors of curl, with the same eigenvalue.

For the real fields $E(r,t)$, $B(r,t)$ we have by superposition

$$
\nabla \times E(r,t) = \pm k E(r,t), \qquad \nabla \times B(r,t) = \pm k B(r,t).
\tag{3.56}
$$

The fact that for self-dual beams the fields are eigenvectors of curl has consequence for chirality, as we shall see in Chapter 5. It will be shown there that self-dual fields are necessarily chiral, and in fact have maximal chirality.

3.7   TM+iTE ELECTROMAGNETIC BEAMS

The self-dual TM+iTE beams are based on the complex vector potential $A = A_{TM} + i A_{TE}$ (Lekner 2001 [2], 2004 [5]) where from (3.12) $A_{TM} = A_0 [0, 0, \psi] = A_0 (0, 0, \psi)$ and

$$A_{TE} = (ik)^{-1} \nabla \times A_{TM} = (ik)^{-1} B_{TM} = A_0 [\partial_y, -\partial_x, 0] \psi = A_0 (\rho^{-1} \partial_\phi, -\partial_\rho, 0) \psi. \quad (3.57)$$

Thus the complex vector potential for TM+iTE beams is, in general,

$$A = A_{TM} + i A_{TE} = k^{-1} A_0 [\partial_y, -\partial_x, k] \psi = k^{-1} A_0 (\rho^{-1} \partial_\phi, -\partial_\rho, k) \psi. \quad (3.58)$$

With azimuthal dependence of ψ in the factor $e^{im\phi}$, the complex field amplitudes are

$$B = \nabla \times A = k^{-1} A_0 (imk\rho^{-1} + \partial_\rho \partial_z, im\rho^{-1} \partial_z - k\partial_\rho, \partial_z^2 + k^2) \psi, \qquad E = iB \quad (3.59)$$

(TM-iTE beams have $E = -iB$.)

The energy and momentum densities are found in terms of the complex magnetic amplitude from (3.53):

$$u(r) = \frac{1}{8\pi} B \cdot B^* = \frac{1}{8\pi} \left\{ |B_\rho|^2 + |B_\phi|^2 + |B_z|^2 \right\}, \qquad c p(r) = \frac{i}{8\pi} B \times B^*. \quad (3.60)$$

The energy, momentum and angular momentum per unit length of TM+iTE beams are evaluated in Lekner 2004 [5] as

$$\begin{bmatrix} U' \\ cP_z' \\ cJ_z' \end{bmatrix} = \frac{A_0^2}{2k} \int_0^k d\kappa \, \kappa \, |f(k,\kappa)|^2 \begin{bmatrix} k \\ q \\ m + \kappa^2/2qk \end{bmatrix} \qquad \text{(TM+iTE)}. \quad (3.61)$$

The methods of evaluation are those developed in Section 3.3, depending on the use of the Bessel function recurrence formulae, the Hankel inversion formula (3.27), and so on. However, the angular momentum contains, as well as $\rho J_m(\kappa\rho) J_m(\kappa'\rho)$, integrals over $\rho^2 J_m(\kappa\rho) J_{m+1}(\kappa'\rho)$ and $\rho^2 J_{m+1}(\kappa\rho) J_m(\kappa'\rho)$. These are even more singular than that leading to the Hankel inversion formula, but symmetric combinations which arise in the evaluation of J_z' are tractable (Appendix 3A). More detail of this kind of calculation may be found in Chapter 4, where we explicitly evaluate J_z' for self-dual beams which are circularly polarized on the beam axis. Note the extra term in the wavenumber integrand, $m + \kappa^2/2qk$ instead of just m as it was for the TM and TE beams. Such a term also arises for the 'CP' beam of Section 4.5, but as $m + 1 + \kappa^2/2qk$. We defer discussion of its physical origin till that Section.

The energy and z component of momentum per unit length of beam, given in (3.61), are just twice those found in Section 3.3. It turns out that when $m = 0$ the energy density and z component of momentum density for the TM+iTE beam are twice the time-averaged densities for the TM or TE beams (based on the same wavefunction). For non-zero m there are additional terms, but these integrate to zero in transverse section, as discussed in Appendix 3B.

3.8   TM+iTE BEAMS-BASED ON ψ_0 AND ψ_1

In contrast to the TM and TE beams of Sections 3.4 and 3.5, we need only to consider the magnetic field, since the complex field amplitudes are related by $E = \pm i\,B$. (We shall take the plus sign in the following.)

For the TM+iTE beam based on ψ_0, the complex field amplitudes in the focal plane $z = 0$ are

$$B_\rho = -2i\rho^{-1} j_2\,(k\rho)\,, \qquad B_\phi = 2\rho^{-1} J_2\,(k\rho)$$
$$B_z = 2\rho^{-3}\left\{2\rho J_0\,(k\rho) + \left[(k\rho)^2 - 4\right] J_1\,(k\rho)\right\}. \qquad (3.62)$$

Comparison with (3.43) and (3.44) for the TM beam shows that the complex amplitudes are related by

$$\left(B_\rho,\ B_\phi,\ B_z\right)_{TM+iTE} = \left(-iE_\rho,\ B_\phi,\ -iE_z\right)_{TM}. \qquad (3.63)$$

The relations (3.63) are in accord with the fact that the TM magnetic field has one non-zero (azimuthal) component, and the $m = 0$ electric azimuthal component is zero; thus $\boldsymbol{B}_{TM+iTE} = \boldsymbol{B}_{TM} + i\,\boldsymbol{E}_{TM}$.

From (3.62) we see that the real magnetic field contains the time-dependence $\sin\omega t$, $\cos\omega t$, $\cos\omega t$ in the radial, azimuthal and longitudinal components, respectively. The radial component is in phase quadrature to the other two (in the focal plane). These three components are shown in Figure 3.7, at their maximum values. The electric field components are in phase quadrature to the magnetic components, since the complex amplitudes are related by $E = i\,B$.

Next we look at the energy, momentum and angular momentum densities. These are most easily expressed and calculated in terms of the complex magnetic amplitude, as in (3.60). We are interested in u, cp_z and ckj_z. For the $m = 0$ TM beams the angular momentum is zero, since p_ϕ is zero, but this is no longer true for the TM+iTE beams. The longitudinal component of angular momentum per unit length of the beam, $J'_z = 2\pi \int_0^\infty d\rho\,\rho\,j_z$, is positive. Spatial integration using the Weber–Schafheitlin integral (3.36), or evaluation of the wavenumber integrals in (3.61) gives

$$U' = \frac{1}{2}A_0^2, \qquad cP'_z = \frac{4}{15}A_0^2, \qquad ckJ'_z = \frac{8}{15}A_0^2 \quad (\text{TM+iTE},\ m = 0) \qquad (3.64)$$

The ratio $cP'_z/U' = 8/15$ is the same as for the TM or TE beams based on ψ_0, as given in (3.34). The energy and longitudinal momentum per unit length of the beam are twice those given in (3.34), as discussed in Appendix 3B. Figure 3.8 shows the energy, longitudinal momentum and angular momentum densities for a beam constructed from ψ_0 in the focal plane.

In Appendix 3B we show that the momentum and energy densities for the $m = 0$ TM+iTE beam are just twice the cycle-averaged densities for the TM or TE beams, shown in Figure 3.3 in longitudinal section, so we do not repeat the u, $c\boldsymbol{p}$ plot here. Instead, we show

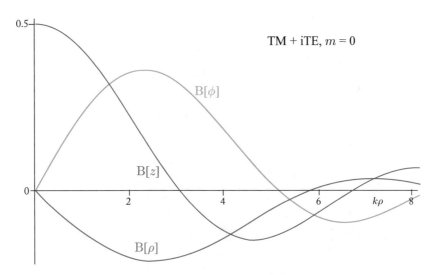

Figure 3.7: The components of **B** of a TM+iTE beam based on ψ_0, in the focal plane. $B_\phi(\rho, 0)$, $B_z(\rho, 0)$ oscillate in phase, and the radial component $B_\rho(\rho, 0)$ is in phase quadrature to them. The ordinate gives the field amplitudes divided by $E_0 = kA_0$.

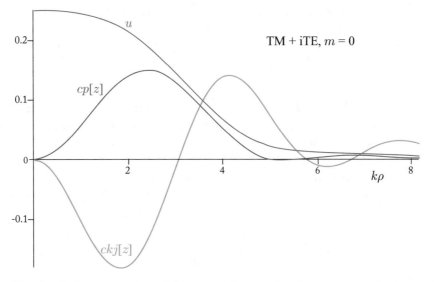

Figure 3.8: The focal plane variation of the energy, longitudinal momentum density and angular momentum densities of the TM+iTE beam based on ψ_0. The ordinate times $E_0^2/8\pi$ ($E_0 = kA_0$) gives u, cp_z and ckj_z. The energy and momentum densities are just twice those for the TM or TE beams based on the same $m = 0$ wave function, shown in Figure 3.2.

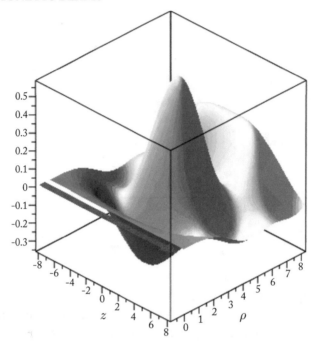

Figure 3.9: The angular momentum integrand ρj_z for the TM+iTE beam based on the proto-beam wavefunction ψ_0, in the neighborhood of the focal region. The arrow gives the direction of beam propagation. Compare with Figure 3.8, which gives the focal plane variation of j_z (not of ρj_z, which is plotted here), that is in the plane normal to the arrow and half-way along it.

the quite complex distribution of the angular momentum integrand ρj_z in Figure 3.9. As also seen in the focal plane plot of Figure 3.8, the angular momentum density j_z has negative as well as positive regions. The net result of the intergration is a positive J_z, as given in (3.64).

We conclude this Section with the TM+iTE beam constructed from ψ_1. The invariant quantities U', P'_z and J'_z (the energy, momentum and angular momentum per unit length of the beam) may be evaluated by spatial integration using the Weber–Schafheitlin integral (3.36), or by evaluation of the wavenumber integrals in (3.61). the results are

$$U' = \frac{1}{3}A_0^2, \quad cP'_z = \frac{16}{105}A_0^2, \quad ckJ'_z = \frac{83}{105}A_0^2 \quad (\text{TM+iTE}, \ m = 1). \tag{3.65}$$

The ratio $cP'_z/U' = 16/35$ is the same as for TM or TE beams based on ψ_1, given in (3.40). The energy and longitudinal momentum per unit length of the beam are twice those given in (3.40), as shown in Appendix 3B.

Figure 3.10 shows the magnetic field components of the TM+iTE beam based on ψ_1. The energy and longitudinal momentum and angular momentum densities for the same beam are

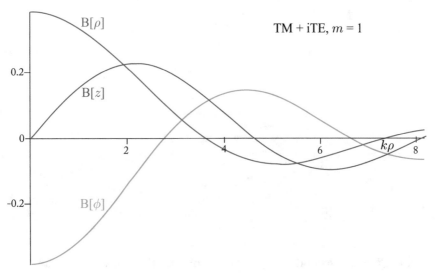

Figure 3.10: The components of **B** of a TM+iTE beam based on ψ_1, in focal plane. The amplitudes $B_\phi\,(\rho,0)$, $B_z\,(\rho,0)$ are real, and the radial component $B_\rho\,(\rho,0)$ is imaginary, after a common phase factor $e^{i\phi}$ has been removed. The ordinate gives the field amplitudes divided by $E_0 = kA_0$.

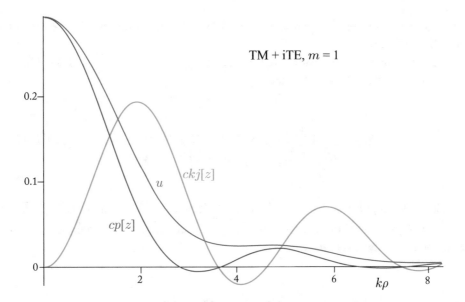

Figure 3.11: The focal plane variation of the energy, longitudinal momentum density and angular momentum densities of the TM+iTE beam based on ψ_1. The ordinate times $E_0^2/8\pi$ $(E_0 = kA_0)$ gives u, cp_z and ckj_z.

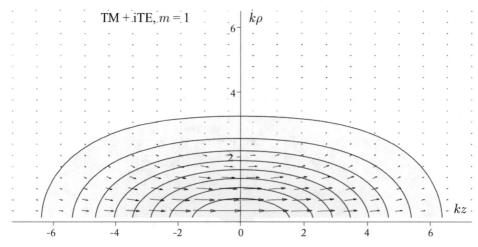

Figure 3.12: Longitudinal section of the density (contours), and the momentum density vector field (arrows) for the TM+iTE beam based on the proto-beam wavefunction ψ_1. The focal region is centered on the origin. A three-dimensional picture is obtained by rotating the figure about the horizontal axis.

shown in Figure 3.11 in the focal plane. Figure 3.12 shows the momentum and energy densities in longitudinal section.

In the next Chapter we consider polarization properties of the beams of this Chapter, and of other kinds of beams to be constructed there.

3A   APPENDIX: TWO SINGULAR INTEGRALS OVER PRODUCTS OF BESSEL FUNCTIONS

We consider the singular integrals $\int_0^\infty d\rho\, \rho^2\, J_m(\kappa\rho)\; J_{m+1}(\kappa'\rho)$ and $\int_0^\infty d\rho\, \rho^2 J_{m+1}(\kappa\rho)$ $J_m(\kappa'\rho)$. As in Appendix B of Lekner 2004 [5], we shall insert a convergence factor, $e^{-\rho^2/a^2}$, and define the integrals as the limit $a \to \infty$ of

$$
\begin{aligned}
U_m\left(a;\kappa,\kappa'\right) &= \int_0^\infty d\rho\, \rho^2\, e^{-\rho^2/a^2}\, J_m\left(\kappa\rho\right)\; J_{m+1}\left(\kappa'\rho\right) \\
V_m\left(a;\kappa,\kappa'\right) &= \int_0^\infty d\rho\, \rho^2\, e^{-\rho^2/a^2}\, J_{m+1}\left(\kappa\rho\right)\; J_m\left(\kappa'\rho\right).
\end{aligned}
\tag{3A.1}
$$

The integrals in (3A.1) are related to Weber's second exponential integral (Watson 1944 [6], Section 13.31)

$$W_m\left(a;\kappa,\kappa'\right) = \int_0^\infty d\rho\, \rho\, e^{-\rho^2/a^2} J_m\left(\kappa\rho\right)\, J_m\left(\kappa'\rho\right)$$

$$= \frac{a^2}{2} e^{-\left(\kappa^2+\kappa'^2\right)a^2/4} I_m\left(\frac{\kappa\kappa'a^2}{2}\right). \tag{3A.2}$$

For example, the relationship between V_m and W_m is

$$V_m\left(a;\kappa,\kappa'\right) = \left(m\kappa^{-1} - \partial_\kappa\right) W_m\left(a;\kappa,\kappa'\right)$$

$$= \frac{a^4}{4} e^{-\left(\kappa^2+\kappa'^2\right)a^2/4} \left\{\kappa I_m\left(\frac{\kappa\kappa'a^2}{2}\right) - \kappa' I_{m+1}\left(\frac{\kappa\kappa'a^2}{2}\right)\right\}. \tag{3A.3}$$

The integral U_m is given by the expression in (3A.3) with κ, κ' interchanged.

In Lekner 2004 [5] it is shown that the limit of W_m as a tends to infinity gives (3.27):

$$\lim_{a\to\infty} W_m\left(a;\kappa,\kappa'\right) = \left(\kappa\kappa'\right)^{-\frac{1}{2}} \delta\left(\kappa - \kappa'\right). \tag{3A.4}$$

The integrals U_m, V_m are more singular in the $a \to \infty$ limit. For example, the leading terms in the asymptotic expansion of (3A.3) as a tends to infinity are

$$V_m\left(a;\kappa,\kappa'\right) \sim \frac{a}{4\kappa\kappa'\sqrt{\pi\kappa\kappa'}} e^{-(\kappa-\kappa')^2a^2/4}$$

$$\left\{a^2\kappa\kappa'\left(\kappa - \kappa'\right) - m^2\left(\kappa - \kappa'\right) + \kappa'\left(2m + 3/4\right) + \kappa/4\right\}. \tag{3A.5}$$

For large a (assumed positive) and fixed κ, the expression (3A.5) has a maximum at $\kappa' = \kappa - \sqrt{2}/a + m/\kappa a^2 + O(a^{-3})$, and a minimum at $\kappa' = \kappa + \sqrt{2}/a + m/\kappa a^2 + O(a^{-3})$. These maxima and minima take the values

$$\pm \left(\frac{2}{\pi e}\right)^{1/2} \frac{a^2}{4\kappa} + \left(\frac{1}{\pi e}\right)^{1/2} \frac{(m+1)a}{4\kappa^2} + O(1). \tag{3A.6}$$

Note that the maxima are proportional to a^2 rather than to a, as is the case for the asymptotic form of W_m. However, various symmetric sums of U_m and V_m do behave like a delta function, because of cancellation of the most singular parts. The simplest is

$$S_m = U_m + V_m = \frac{(\kappa+\kappa')a^4}{4} e^{-\left(\kappa^2+\kappa'^2\right)a^2/4} \left\{I_m\left(\frac{\kappa\kappa'a^2}{2}\right) - I_{m+1}\left(\frac{\kappa\kappa'a^2}{2}\right)\right\}$$

$$= \frac{(2m + 1)(\kappa + \kappa')}{4\kappa\kappa'\sqrt{\pi\kappa\kappa'}} e^{-(\kappa-\kappa')^2a^2/4} \left\{a - \frac{(2m - 1)(2m + 3)}{4a\kappa\kappa'} + O(a^{-3})\right\}. \tag{3A.7}$$

The maximum at $\kappa = \kappa'$ is $(2m + 1) a/2\kappa^2 \sqrt{\pi} + O(a^{-1})$, and has width of order a^{-1}. In other words, $S_m (a; \kappa, \kappa')$ acts like the delta function $\delta(\kappa - \kappa')$ as a tends to infinity. To find the proportionality factor, we integrate (3A.7) over κ' in the neighborhood of κ when a is large. The result is $(2m + 1)\kappa^{-2}$ plus terms of order a^{-2}, so

$$\lim_{a \to \infty} S_m (a; \kappa, \kappa') = (2m + 1)(\kappa\kappa')^{-1} \delta (\kappa - \kappa'). \tag{3A.8}$$

Other symmetric additive combinations of U_m and V_m also behave like delta functions. The one needed in the evaluation of J_z' for the TM+iTE beams as given in (3.61) is

$$\kappa' q U_m + \kappa q' V_m$$
$$= \int_0^\infty d\rho \, \rho^2 \, e^{-\rho^2/a^2} \left\{ \kappa' q J_{m+1} (\kappa\rho) \, J_m (\kappa'\rho) + \kappa q' J_m (\kappa\rho) \, J_{m+1} (\kappa'\rho) \right\}. \tag{3A.9}$$

Analysis along the lines just outlined for S_m gives

$$\lim_{a \to \infty} \left\{ \kappa' q U_m + \kappa q' V_m \right\} = \frac{1}{\kappa q} \left[(2m + 2) k^2 - (2m + 1)\kappa^2 \right] \delta(\kappa - \kappa'). \tag{3A.10}$$

3B   APPENDIX: COMPARISON OF TM OR TE BEAM ENERGY AND MOMENTUM DENSITIES WITH THOSE OF A TM+iTE BEAM

Equations (3.33) and (3.61) give the energy, z component of momentum, and z component of angular momentum per unit length of TM or TE, and TM+iTE beams, respectively. The values for the energy and the z component of momentum per unit length for the latter are twice those of the former. This is no accident, but neither is it trivially true. The fields of TM+iTE beams are superpositions of TM fields and TE fields, in phase quadrature. That does not imply a simple relation between the energy or the momentum densities, except when $m = 0$. We shall show here that when $m = 0$ the energy density and z component of momentum density for the TM+iTE beam are twice the time-averaged densities for the TM or TE beams (based on the same wavefunction). For m non-zero there are additional terms, but these integrate to zero in a transverse section.

We begin with the energy densities. Let u denote the energy density of a TM+iTE beam, and $\bar{u}$ the cycle-averaged energy density of a TM or TE beam, in all cases based on the same beam wavefunction ψ. It is assumed that the azimuthal dependence is in the factor $e^{im\phi}$, and that ψ satisfies the Helmholtz equation. Then we find, after some reduction,

$$\frac{8\pi}{k^2 A_0^2} (u - 2\bar{u}) = \frac{2m}{k\rho} \text{Im} \left\{ (\partial_\rho \psi^*) (\partial_z \psi) - (\partial_\rho \partial_z \psi^*) \psi \right\}. \tag{3B.1}$$

Hence $u = 2\bar{u}$ when $m = 0$. When m is not zero the quantities u and $2\bar{u}$ are not equal, in general, but the energy U' per unit length of a TM+iTE beam is still twice that of a TM or

TE beam. To show this, we need to integrate both sides of (3B.1) over a section of the beams at fixed z, operating with $\int d^2r = 2\pi \int_0^\infty d\rho \, \rho$ (the azimuthal dependence has cancelled out). Integration by parts gives us

$$\int_0^\infty d\rho \, \text{Im} \left\{ (\partial_\rho \psi^*)(\partial_z \psi) - (\partial_\rho \partial_z \psi^*) \psi \right\}$$

$$= \int_0^\infty d\rho \, \text{Im} \left\{ (\partial_\rho \psi^*)(\partial_z \psi) + (\partial_z \psi^*)(\partial_\rho \psi) \right\} + \text{Im} \left[(\partial_z \psi^*) \psi \right]_{\rho=0}. \tag{3B.2}$$

Both terms in the final line are zero: the first because the quantity in braces is real, and the second because for $m \neq 0$ the general beam wavefunction (3.23) is zero on the beam axis.

Next we look at the z component of the momentum; we find, again after reduction,

$$\frac{8\pi}{k^2 A_0^2} (p_z - 2\overline{p_z}) = \frac{2m}{k^2 \rho} \text{Re} \left\{ k^2 (\partial_\rho \psi^*)(\psi) + (\partial_\rho \partial_z \psi^*)(\partial_z \psi) \right\}. \tag{3B.3}$$

Thus, $p_z = 2\overline{p_z}$ for $m = 0$ beams, where p_z is the momentum density of the TM+iTE beam, and $\overline{p_z}$ is the cycle-averaged momentum density of either a TM or a TE beam based on the same wavefunction. When m is not zero the quantities p_z and $2\overline{p_z}$ are not in general equal, but the momentum per unit length of a TM+iTE beam is still twice that of a TM or TE beam. We again integrate both sides of (3B.3) over a section of the beams at fixed z, operating with $\int d^2r = 2\pi \int_0^\infty d\rho \, \rho$. The first term is proportional to

$$\int_0^\infty d\rho \left\{ (\partial_\rho \psi^*) \psi + \psi^* \partial_\rho \psi \right\} = \int_0^\infty d\rho \, \partial_\rho (\psi^* \psi) = 0. \tag{3B.4}$$

(For $m \neq 0$ the general beam wavefunction (3.23) is zero on the beam axis, and it is also zero at infinity.) The second term in (3B.3) integrates similarly to $(\partial_z \psi^*)(\partial_z \psi)$, to be evaluated at the limits $\rho = 0, \infty$. We can again verify from (3.23) that this quantity is zero at the limits stated.

3.11 REFERENCES

[1] Davis, L. W. and Patsakos, G. 1981. TM and TE electromagnetic beams in free space, *Optics Letters*, 6:22–23. DOI: 10.1364/ol.6.000022. 40

[2] Lekner, J. 2001. TM, TE and "TEM" beam modes: Exact solutions and their problems, *Journal of Optics A: Pure and Applied Optics*, 3:407–412. DOI: 10.1088/1464-4258/3/5/314. 53

[3] Lekner, J. 2002. Phase and transport velocities in particle and electromagnetic beams, *Journal of Optics A: Pure and Applied Optics*, 4:491–499. DOI: 10.1088/1464-4258/4/5/301. 52

[4] Lekner, J. 2003. Polarization of tightly focused laser beams, *Journal of Optics A: Pure and Applied Optics*, 5:6–14. DOI: 10.1088/1464-4258/5/1/302.

[5] Lekner, J. 2004. Invariants of three types of generalized Bessel beams, *Journal of Optics A: Pure and Applied Optics*, 6:837–843. DOI: 10.1088/1464-4258/6/9/004. 42, 43, 44, 53, 58, 59

[6] Watson, G. N. 1944. *Theory of Bessel Functions*, Cambridge University Press. 42, 43, 45, 59

CHAPTER 4

Polarization

4.1   POLARIZATION OF ELECTROMAGNETIC FIELDS

We consider monochromatic fields, with complex time-dependence $e^{-i\omega t}$. The angular frequency is $\omega = ck$. The complex field amplitudes are $\boldsymbol{E}(\boldsymbol{r})e^{-i\omega t}$, $\boldsymbol{B}(\boldsymbol{r})e^{-i\omega t}$. Real electric and magnetic fields are obtained by taking real or imaginary parts of the complex field amplitudes. Taking the real part gives, with $\boldsymbol{E}(\boldsymbol{r}) = \boldsymbol{E}_r(\boldsymbol{r}) + i\boldsymbol{E}_i(\boldsymbol{r})$,

$$\boldsymbol{E}(\boldsymbol{r},t) = Re\left\{\boldsymbol{E}(\boldsymbol{r})e^{-i\omega t}\right\} = Re\left\{(\boldsymbol{E}_r + i\boldsymbol{E}_i)e^{-i\omega t}\right\} = \boldsymbol{E}_r\cos\omega t + \boldsymbol{E}_i\sin\omega t. \qquad (4.1)$$

We shall discuss primarily electric polarization. Magnetic polarization is of lesser interest experimentally, but the formalism is the same. At a fixed point in space the endpoint of the vector $\boldsymbol{E}(\boldsymbol{r},t)$ describes an ellipse in time $2\pi/\omega$ (see for example Born and Wolf 1999, Section 1.4.3 [8]): one can write

$$\boldsymbol{E}_r + i\boldsymbol{E}_i = (\boldsymbol{E}_1 + i\boldsymbol{E}_2)e^{i\gamma}, \qquad \tan 2\gamma = \frac{2\boldsymbol{E}_r \cdot \boldsymbol{E}_i}{E_r^2 - E_i^2}. \qquad (4.2)$$

The real field is then

$$\boldsymbol{E}(\boldsymbol{r},t) = Re\left\{(\boldsymbol{E}_1 + i\boldsymbol{E}_2)e^{i\gamma - i\omega t}\right\} = \boldsymbol{E}_1\cos(\omega t - \gamma) + \boldsymbol{E}_2\sin(\omega t - \gamma). \qquad (4.3)$$

From the first relation in (4.2) we have

$$\boldsymbol{E}_1 = \boldsymbol{E}_r\cos\gamma + \boldsymbol{E}_i\sin\gamma, \qquad \boldsymbol{E}_2 = \boldsymbol{E}_i\cos\gamma - \boldsymbol{E}_r\sin\gamma. \qquad (4.4)$$

Then when the second relation in (4.2) holds the transformed components are orthogonal: $\boldsymbol{E}_1 \cdot \boldsymbol{E}_2 = 0$. Hence from (4.3) the electric field vector endpoint describes an ellipse with semiaxes E_1, E_2. These have the squared magnitudes given by

$$\begin{pmatrix} E_1^2 \\ E_2^2 \end{pmatrix} = \frac{1}{2}\left[E_r^2 + E_i^2 \pm \sqrt{(E_r^2 - E_i^2)^2 + 4(\boldsymbol{E}_r \cdot \boldsymbol{E}_i)^2} \right]. \qquad (4.5)$$

(We have chosen the plus sign for E_1, and consequently that $E_1 \geq E_2$.) Thus, in general, monochromatic electromagnetic waves have fixed elliptical polarization at any point in space. Special cases are linear and circular polarization, in which the endpoint of the electric field vector moves on a line, and on a circle.

For *linear* polarization of the electric field the real and imaginary parts of the complex vector $E = E_r + i E_i$ are collinear:

$$E_r^2 E_i^2 - (E_r \cdot E_i)^2 = 0 \qquad \text{(linear polarization).} \qquad (4.6)$$

Note that collinearity includes the possibility that one of E_r, E_i is zero. The condition for E_r and E_i to be collinear can also be written as $E_r \times E_i = 0$ (equivalent to $E^* \times E = 2i\, E_r \times E_i = 0$). The square of this relation gives (4.6). ($E_r \times E_i = 0$ is a vector equation with three components, but is equivalent to one scalar relation: the vectors E_r, E_i define a plane, and the non-zero component of their vector product is perpendicular to this plane.)

For *circular* polarization the real and imaginary parts of the complex amplitude E are perpendicular and equal in magnitude:

$$\{E_r \cdot E_i = 0 \quad \text{and} \quad E_r^2 = E_i^2\} \qquad \text{(circular polarization).} \qquad (4.7)$$

In between the two limiting polarizations lie all possible elliptical polarizations, in which the endpoint of the real vector $E(r,t)$ given in (4.3) describes an ellipse in time $2\pi/\omega$ at each point in space. Note that for monochromatic fields the polarization is a function only of the position.

Various polarization measures are possible: for example we can use the eccentricity e of the polarization ellipse, defined by

$$e^2 = 1 - E_2^2/E_1^2 = \frac{2\sqrt{\left(E_r^2 - E_i^2\right)^2 + 4(E_r \cdot E_i)^2}}{E_r^2 + E_i^2 + \sqrt{\left(E_r^2 - E_i^2\right)^2 + 4(E_r \cdot E_i)^2}}. \qquad (4.8)$$

The eccentricity is zero for circular polarization, and unity for linear polarization. Hurwitz (1945) [9] used the parameter

$$S = \frac{2E_1 E_2}{E_1^2 + E_2^2} = \frac{2\sqrt{E_r^2 E_i^2 - (E_r \cdot E_i)^2}}{E_r^2 + E_i^2}. \qquad (4.9)$$

The Hurwitz parameter is zero for linear polarization, and unity for circular polarization.

A measure of the degree of linear polarization used in Lekner 2003 [11] is the ratio

$$\Lambda(r) = \frac{|E^2(r)|}{|E(r)|^2} = \frac{|E(r) \cdot E(r)|}{E(r) \cdot E(r)^*} = \frac{|(E_r + i E_i)^2|}{|E_r + i E_i|^2}$$

$$= \frac{\left[\left(E_r^2 - E_i^2\right)^2 + 4(E_r \cdot E_i)^2\right]^{\frac{1}{2}}}{E_r^2 + E_i^2} = \frac{E_1^2 - E_2^2}{E_1^2 + E_2^2}. \qquad (4.10)$$

It is clear from (4.10) that $\Lambda(r) = 1$ on surfaces where E_r, E_i are collinear and the field is linearly polarized, while $\Lambda(r) = 0$ on curves where the conditions (4.7) for circular polarization are satisfied. The same measure applies to B, which in the case of self-dual beams with $E = \pm i B$ has the same polarization as E, since then $B_r = \pm E_i$, $B_i = \mp E_r$.

These three polarization measures are related by $e^2 = 2\Lambda/(1 + \Lambda)$, $S^2 = 1 - \Lambda^2$. All three measures give the degree of linear or circular polarization, but Λ has the advantage of being simply expressed in terms of the original complex field amplitude $E(r)$, and of making explicit the fact that circular polarization corresponds to nilpotency of the complex field, that is to $E^2(r) = 0$. It is clear from the definitions of e, S and Λ that the polarization measures are unchanged by an overall change of either phase or amplitude of the field.

The three measures just discussed do not give information about the orientation of the polarization ellipse. The full information lies in the fields E_r, E_i, or in the Stokes parameters (Born and Wolf 1999, Sections 1.4.2 and 10.8.3 [8])

$$
\begin{aligned}
s_0 &= E_r^2 + E_i^2, & s_1 &= E_r^2 - E_i^2 \\
s_2 &= 2E_r \cdot E_i, & s_3 &= 2\sqrt{E_r^2 E_i^2 - (E_r \cdot E_i)^2}.
\end{aligned}
\tag{4.11}
$$

The three scalar polarization measures are given in terms of the Stokes parameters s_0, s_3 by

$$
S = \frac{s_3}{s_0}, \qquad \Lambda = \sqrt{1 - S^2}, \qquad e^2 = \frac{2\Lambda}{1 + \Lambda}.
\tag{4.12}
$$

An alternative vector polarization measure is Lindell's (1992) [15] vector (I have changed the sign)

$$
P(E) = \frac{E^* \times E}{i E^* \cdot E} = \frac{2E_r \times E_i}{E_r^2 + E_i^2}.
\tag{4.13}
$$

This vector is normal to the plane defined by E_r, E_i (the plane of the ellipse traced out by the endpoint of the real electric field, in general). The vector magnitude is zero for linear polarization, and unity for circular polarization.

Nye and Hajnal (1987) [17] and Nye (1999) [19] have studied and classified the polarization of electromagnetic fields geometrically. The location of circular polarization of monochromatic fields is specified by the two conditions (4.7), namely $E_r \cdot E_i = 0$ and $E_r^2 = E_i^2$. Each condition determines a surface in space, and the two surfaces intersect on curves C on which the electric field is circularly polarized. The location of linear polarization is determined by one condition, $E_r^2 E_i^2 = (E_r \cdot E_i)^2$, which determines a surface S in space. The curves C cannot cross an S surface, except at a point where the field is zero.

Examples of these geometric aspects will be shown in the following Sections.

4.2   POLARIZATION PROPERTIES OF TM AND TE BEAMS

As the simplest example of polarization, consider the TM (transverse magnetic) monochromatic beams discussed in Sections 3.2–3.5. For these beams the vector potential is along the beam axis, $\boldsymbol{A} = [0, 0, \psi] = (0, 0, \psi)$.

We shall assume first that the beam wavefunction does not depend on the azimuthal angle: we set $\psi = \psi(\rho, z)$, $m = 0$ in (2.14). Then the complex magnetic amplitude $\boldsymbol{B} = \nabla \times \boldsymbol{A} = (0, -\partial_\rho \psi, 0)$ is everywhere azimuthal. The *magnetic field is linearly polarized*, since there is only one non-zero field component and the real and imaginary parts of the amplitude $\boldsymbol{B}$ are collinear. The measure of linear polarization Λ_B is everywhere unity. The magnetic field lines are circles, with centers on the beam axis. The electric and magnetic fields are everywhere perpendicular.

The electric complex amplitude has radial and longitudinal components, $\boldsymbol{E} = i k^{-1} (\partial_\rho \partial_z, 0, \partial_z^2 + k^2)\psi$, and the electric field is elliptically polarized in general, as discussed below. We shall take as an example $\psi = \psi_0$, the proto-beam wavefunction. Then in the focal plane the complex electric amplitude is, from the formulae (3.44),

$$\boldsymbol{E}(\rho, 0) \sim \left(j_2(k\rho), \ 0, \ i \left\{ 2(k\rho)^{-2} [k\rho J_0(k\rho) - 2J_1(k\rho)] + J_1(k\rho) \right\} \right). \tag{4.14}$$

(Common factors have been omitted; they do not affect polarization measures.) In the focal plane the real part of the amplitude is radial, and the imaginary part is longitudinal. The polarization measure is therefore, from (4.10),

$$\Lambda_E = \frac{|E_r^2 - E_i^2|}{E_r^2 + E_i^2}, \quad E_r \sim j_2(k\rho), \quad E_i \sim 2(k\rho)^{-2} [k\rho J_0(k\rho) - 2J_1(k\rho)] + J_1(k\rho). \tag{4.15}$$

The vectors $\boldsymbol{E}_r$, $\boldsymbol{E}_i$ are radial and longitudinal, respectively. There will be rings of circular polarization when $E_r^2 = E_i^2$, and surfaces of linear polarization intersecting the focal plane at circles when either of E_r, E_i is zero. Figure 4.1 shows focal plane variation of the radial and longitudinal field components, and the polarization measure Λ_E, based on the proto-beam wavefunction ψ_0. Note the rapid variation in polarization, from linear at the center of the focal region, down to circular, up again to linear, and so on. The energy density of the beam is still substantial for the first three of these changes (see Figure 3.2). Where the electric field is linearly polarized, it is either longitudinal (when $E_\rho = 0$, as happens at the focal center and on the beam axis) or radial (when $E_z = 0$). When circularly polarized, the endpoint of the electric field vector moves on circles which lie in planes through the beam axis.

TM beams with $m = 0$ have linear electric polarization on the beam axis, $\Lambda_E = 1$. This is because, from (3.16), the radial component of the electric amplitude is proportional to $\partial_\rho \partial_z \psi$, and for $m = 0$ is thus proportional to an integral over $J_1(\kappa\rho)$, which is zero on the beam axis. This leaves one non-zero component of the complex electric field, E_z, and the real and imaginary parts of $\boldsymbol{E}$ are necessarily parallel. The magnetic field is linearly polarized everywhere when $m = 0$, because only the azimuthal field component is non-zero.

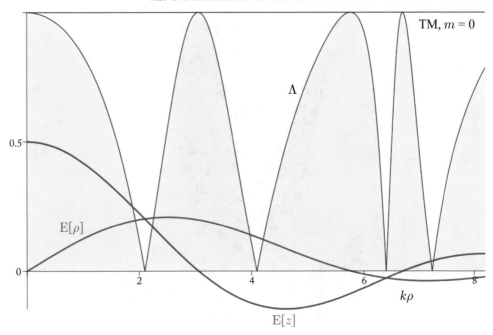

Figure 4.1: Electric field components and polarization of a TM beam based on the proto-beam wavefunction ψ_0, in the focal plane. For $m = 0$ the azimuthal field E_ϕ is zero. Shown are the radial and longitudinal field components (which are in phase quadrature), and the polarization measure Λ_E. Linear polarization ($\Lambda_E = 1$) occurs when one of the perpendicular electric field components is zero. Circular polarization ($\Lambda_E = 0$) results when the perpendicular electric field components are equal in magnitude. Properties of the same TM beam appear in Figures 3.1, 3.2, and 3.3. Figure 3.2 shows that the energy density in the focal plane has dropped to 10% of its peak value at about $k\rho = 5$.

The description of polarization, which has so far been about the TM beam, applies to its dual the TE beam, with substitutions $\boldsymbol{E} \to \boldsymbol{B}$, $\boldsymbol{B} \to -\boldsymbol{E}$. This duality of course holds also for the non-zero m TM and TE beams, to be considered next.

For beam wavefunctions of the form (3.13) and $\boldsymbol{A}_{TM} = A_0 \, (0, \, 0, \psi)$ the magnetic and electric complex field amplitudes are, from (3.14) and (3.15),

$$\boldsymbol{B}\,(\boldsymbol{r}) = A_0 \left(im\rho^{-1}, \, -\partial_\rho, \, 0\right) \psi,$$
$$\boldsymbol{E}\,(\boldsymbol{r}) = A_0 k^{-1} \left(i\partial_\rho\partial_z, \, -m\rho^{-1}\partial_z, \, i\left[\partial_z^2 + k^2\right]\right) \psi. \tag{4.16}$$

The magnetic field now has two non-zero components in general; it is linearly polarized only at the zeros of either component, and circularly polarized when the two components are in phase quadrature and have equal amplitude. The polarization measure $\Lambda_B = |\boldsymbol{B} \cdot \boldsymbol{B}|/\boldsymbol{B} \cdot \boldsymbol{B}^*$ is

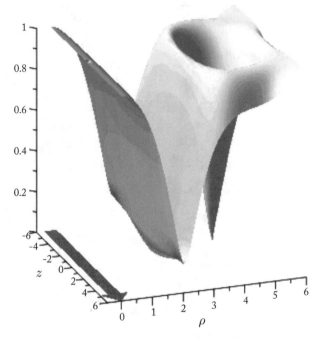

Figure 4.2: The polarization measure Λ_E for the TM $m = 0$ beam based on ψ_0. $\Lambda_E = 1$ (corresponding to linear electric polarization) on the beam axis $\rho = 0$, as shown in the text. Also visible is an annular surface centered on the beam axis on which the electric field is linearly polarized. Inside this surface is a ring on which the polarization is exactly circular (compare Figure 4.1). The arrow gives the direction of beam propagation.

independent of the azimuthal angle ϕ, since this enters as the phase factor $e^{im\phi}$ in all the field components.

We shall use the beam wavefunction ψ_1 of Section 2.8, as we did in Section 3.5. For the $m = 1$ TM beam the non-zero components are B_ρ, B_ϕ and E_ρ, E_ϕ, E_z. The focal plane variation of these field components was shown in Figure 3.4. These field components give the magnetic polarization shown in Figure 4.3 and the electric polarization shown in Figure 4.4.

We saw that when $m = 0$ the electric polarization of TM beams is linear on the beam axis, and that the magnetic field is linearly polarized everywhere. For $m = 1$ both the electric and the magnetic polarizations are circular on the beam axis, as we shall now show. We set $m = 1$ in the general wavefunction (3.13), and expand in powers of ρ:

$$\psi_1 = \frac{\rho}{2}e^{i\phi} \int_0^k d\kappa \, \kappa \, f(k,\kappa) \, e^{iqz} + O(\rho^3). \tag{4.17}$$

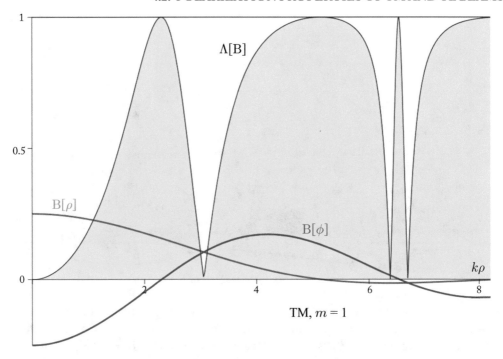

Figure 4.3: Focal plane magnetic field components and magnetic polarization of a TM beam based on the proto-beam wavefunction ψ_1. B_z is zero for TM beams. Shown are the radial and azimuthal field components (which are in phase quadrature), and the polarization measure Λ_B. Linear polarization ($\Lambda_B = 1$) occurs when one of the perpendicular magnetic field components is zero. Circular polarization ($\Lambda_B = 0$) results when the in-quadrature magnetic field components are equal in magnitude. Properties of the same TM beam appear in Figures 3.4, 3.5, and 3.6. Figure 3.5 shows that the energy density in the focal plane has dropped to 10% of its peak value at about $k\rho = 5.8$.

Hence the magnetic field complex amplitude on the beam axis $\rho = 0$ is, from (4.16),

$$\boldsymbol{B}\left(0,z\right) = \frac{A_0}{2}e^{i\phi}\int_0^k d\kappa\,\kappa\,f(k,\kappa)\,e^{iqz}\,(i,\,-1,\,0)\,. \tag{4.18}$$

Likewise from (4.16), the electric complex amplitude on the beam axis is

$$\boldsymbol{E}\left(0,z\right) = \frac{A_0}{2}e^{i\phi}k^{-1}\int_0^k d\kappa\,\kappa\,q\,f(k,\kappa)\,e^{iqz}\,(-1,\,-i,\,0)\,. \tag{4.19}$$

Thus the radial and azimuthal components of the complex magnetic amplitude are in phase quadrature, and equal in magnitude. The same is true for the electric amplitude. These are the conditions for circular polarization, as given in (4.7).

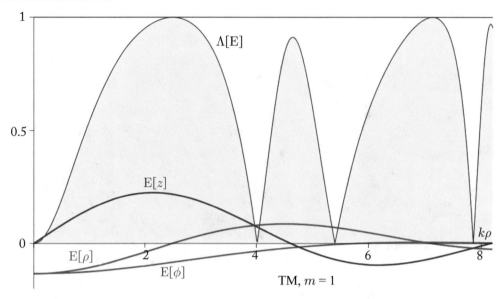

Figure 4.4: Focal plane electric field components and electric polarization measure Λ_E of a TM beam based on the proto-beam wavefunction ψ_1. Shown are the radial, azimuthal and longitudinal field components (E_ρ is in phase quadrature to E_ϕ, E_z). Linear polarization ($\Lambda_E = 1$) occurs when E_ρ is zero, since this leaves the in-phase components E_ϕ, E_z, with collinear real and imaginary parts. There is circular polarization ($\Lambda_E = 0$) on the beam axis, as shown in the text. Properties of the same TM beam appear in Figures 3.4, 3.5, and 3.6. Figure 3.5 shows that the energy density in the focal plane has dropped to 10% of its peak value at about $k\rho = 5.8$.

Figure 4.4 shows focal plane electric field components and the polarization measure Λ_E for the TM beam based on ψ_1. We see that $\Lambda_E = 1$ (linear polarization) when the radial field component is zero. This is because, from (3.48), the complex amplitudes E_ϕ, E_z carry the same phase factor $i e^{i\phi}$, so $\Lambda_E = |\mathbf{E} \cdot \mathbf{E}|/\mathbf{E} \cdot \mathbf{E}^* = \left| E_\phi^2 + E_z^2 \right| / \left(\left| E_\phi \right|^2 + \left| E_z \right|^2 \right) = 1$.

To conclude this Section, we show the longitudinal variation of the polarization of a TM beam based on ψ_1 within the focal region. Figures 4.5 and 4.6 show Λ_B and Λ_E, respectively.

4.3   POLARIZATION IN SELF-DUAL TM+iTE BEAMS

In Section 3.7 we considered the self-dual TM+iTE beam, with vector potential

$$\mathbf{A} = A_0 k^{-1} \left[\partial_y, -\partial_x, k \right] \psi = A_0 k^{-1} (\rho^{-1} \partial_\phi, -\partial_\rho, k) \psi. \tag{4.20}$$

With azimuthal dependence of ψ in the factor $e^{im\phi}$, the complex fields are

$$\mathbf{B} = \nabla \times \mathbf{A} = A_0 k^{-1} \left(\partial_\rho \partial_z + imk\rho^{-1}, -k\partial_\rho + im\rho^{-1}\partial_z, \partial_z^2 + k^2 \right) \psi, \quad \mathbf{E} = i\mathbf{B}. \tag{4.21}$$

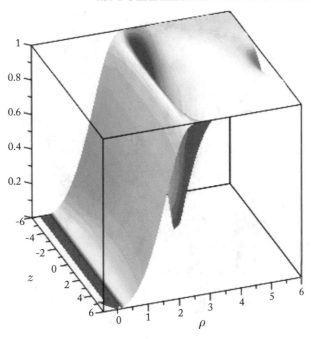

Figure 4.5: The polarization measure Λ_B for the TM $m = 1$ beam based on ψ_1. $\Lambda_B = 0$ (corresponding to circular magnetic polarization) on the beam axis $\rho = 0$, as shown in the text. Also visible is an annular surface centered on the beam axis on which the magnetic field is linearly polarized. Inside this surface is a ring on which the polarization is exactly circular (compare Figure 4.3). The arrow gives the direction of beam propagation.

For self-dual beams the electric and magnetic fields have the same polarization measure Λ at a given point in space; they are elliptically polarized in general. Because $\boldsymbol{E} = \pm i \boldsymbol{B}$ the electric and magnetic fields are in phase quadrature, but are otherwise identical. We note some other general properties in the paragraphs below.

When $m = 0$ the fields are purely longitudinal on the beam axis $\rho = 0$, since the derivative of $J_0(\kappa\rho)$ with respect to ρ is proportional to $J_1(\kappa\rho)$ and is thus zero on the beam axis. Hence TM+iTE beams with $m = 0$ are all linearly polarized on the axis, on which only the z component is non-zero.

For $m = 1$ the wavefunction integrand contains $J_1(\kappa\rho)$, with leading term $\kappa\rho/2$. Thus the magnetic field complex amplitude on the beam axis $\rho = 0$ is, from (4.21),

$$\boldsymbol{B}\,(0, z) = \frac{A_0}{2k} e^{i\phi} \int_0^k d\kappa\, \kappa\, f\,(k, \kappa)\, e^{iqz}\, (iq + ik, -q - k, 0)\,. \qquad (4.22)$$

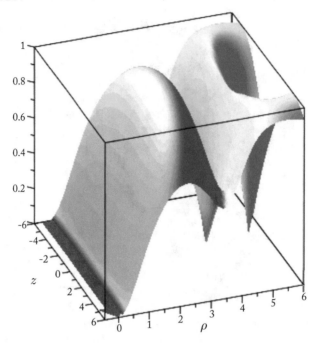

Figure 4.6: The polarization measure Λ_E for the TM $m = 1$ beam based on ψ_1. $\Lambda_E = 0$ (corresponding to circular electric polarization) on the beam axis $\rho = 0$, as shown in the text. Also visible is an annular surface centered on the beam axis on which the electric field is linearly polarized, and two rings on which the polarization is exactly circular (compare Figure 4.4). The arrow gives the direction of beam propagation.

Hence TM+iTE beams with $m = 1$ have the transverse field components equal in magnitude and in phase quadrature on the beam axis. The longitudinal component is zero when $\rho = 0$, so the polarization on the axis is circular, $\Lambda_B = 0 = \Lambda_E$.

Figures 4.7, 4.8, and 4.9, 4.10 show the focal plane polarization measure Λ for TM+iTE beams with $m = 0$ and $m = 1$, respectively. The electric and magnetic fields have the same Λ, since TM+iTE beams are self-dual. The field components for the TM+iTE beam based on ψ_0 were shown in Figure 3.7, and energy and momentum densities for the TM+iTE beam based on ψ_1 were shown in Figures 3.10 and 3.11.

We note that in both pairs of Figures full linear polarization ($\Lambda = 1$) coincides with a zero of p_z. This is because, in the focal plane, the real and imaginary parts of the complex magnetic amplitude take the form

$$\boldsymbol{B}_r = (0, B_\phi, B_z), \quad \boldsymbol{B}_i = (B_\rho, 0, 0) \quad \text{[where } B_\rho, B_\phi, \text{ and } B_z \text{ are real functions].} \quad (4.23)$$

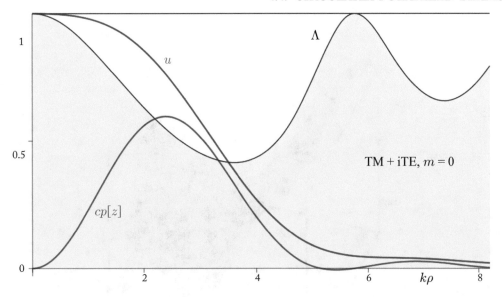

Figure 4.7: The focal plane polarization measure Λ for the TM+iTE beam based on ψ_0. Note full linear polarization at the origin. This persists on the beam axis, as discussed in the text. Also shown are the energy density (normalized to unity at the origin, the center of the focal region), and c times the z component of the momentum density (multiplied by the same factor as the energy density).

Equation (4.23) is valid for the TM+iTE beam based on ψ_0; the same form holds for the $m = 1$ TM+iTE beam based on ψ_1 if we take out the common factor $e^{i\phi}$ from the field components. From (3.53) we have $4\pi c p_z = (\boldsymbol{B}_r \times \boldsymbol{B}_i)_z = B_\rho B_\phi$. In both cases the zeros of B_ρ leave two in-phase components, and thus give linear polarization. On the other hand, a zero of B_ϕ also gives zero p_z, but the field then has two components in quadrature, and is not linearly polarized.

4.4   'CIRCULARLY POLARIZED' BEAMS

Although 'circularly polarized' beams are (in vector potential and electric and magnetic fields) superpositions of beams polarized linearly along perpendicular transverse axes and in phase quadrature, the resultant energy and momentum densities and polarization measures are *not* superpositions, because all of these quantities are bilinear in the field amplitudes.

The vector potential corresponding to that of a self-dual 'circularly polarized' beam is given by Lekner (2003, 2016) [11, 13]. We shall give expressions in both Cartesian $[x, y, z]$ and polar (ρ, ϕ, z) coordinates; it is assumed that the ϕ dependence of ψ is in the factor $e^{im\phi}$, as in (3.13)

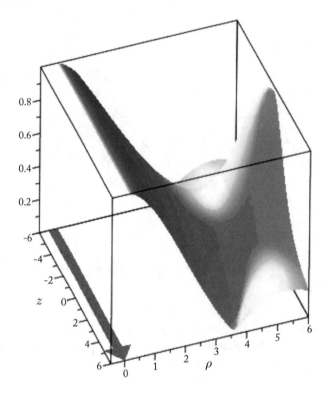

Figure 4.8: The polarization measure Λ for the $m = 0$ TM+iTE beam based on ψ_0. $\Lambda = 1$ (corresponding to linear polarization) on the beam axis $\rho = 0$, as shown in the text. Also visible is a focal plane ring centered on the beam axis on which the electric and magnetic fields are linearly polarized (compare Figure 4.7). Away form the focal plane there are rings of circular polarization, $\Lambda = 0$.

which we repeat here:

$$\psi(r) = e^{im\phi} \int_0^k d\kappa\, f(k,\kappa)\, e^{iqz} J_m(\kappa\rho), \qquad q = \sqrt{k^2 - \kappa^2}. \qquad (4.24)$$

In that case we have

$$\partial_x = \cos\phi\partial_\rho - \rho^{-1}\sin\phi\partial_\phi \to \cos\phi\partial_\rho - im\rho^{-1}\sin\phi$$
$$\partial_y = \sin\phi\partial_\rho + \rho^{-1}\cos\phi\partial_\phi \to \sin\phi\partial_\rho + im\rho^{-1}\cos\phi \qquad (4.25)$$
$$\partial_x + i\partial_y = e^{i\phi}\left(\partial_\rho + i\rho^{-1}\partial_\phi\right) \to e^{i\phi}(\partial_\rho - m\rho^{-1}).$$

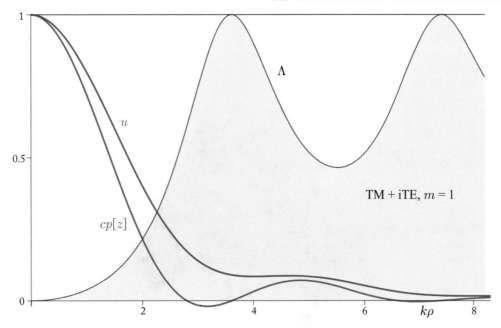

Figure 4.9: The focal plane polarization measure Λ for the TM+iTE beam based on ψ_1. Note exactly circular polarization at the origin (which continues on the entire z axis, as discussed in the text), and the circles of full linear polarization surrounding it. Also shown are the energy density (normalized to unity at the origin, the center of the focal region), and c times the z component of the momentum density (multiplied by the same factor as the energy density).

Consider first the vector potential $A_1 = k^{-1} E_0 [-i, 1, 0] \psi = k^{-1} E_0 e^{i\phi} (-i, 1, 0) \psi$. The fields derived from this vector potential are

$$
\begin{aligned}
B_1 &= k^{-1} E_0 [-\partial_z, -i\partial_z, \partial_x + i\partial_y, 0] \psi = k^{-1} E_0 e^{i\phi} \left(-\partial_z, -i\partial_z, \partial_\rho - m\rho^{-1}\right) \psi \\
E_1 &= k^{-2} E_0 \left[\partial_x (\partial_x + i\partial_y) + k^2, \partial_y (\partial_x + i\partial_y) + ik^2, (\partial_x + i\partial_y)\partial_z\right] \psi \qquad (4.26) \\
&= k^{-2} E_0 e^{i\phi} \left(\partial_\rho^2 - m\rho^{-1}\partial_\rho + k^2 - m\rho^{-2}, (m+1)\rho^{-1}\partial_\rho + k^2 - m(m+1)\rho^{-2},\right. \\
&\quad \left. \partial_\rho\partial_z - m\rho^{-1}\partial_z\right) \psi.
\end{aligned}
$$

In the plane wave limit $\psi \to e^{ikz}$, $B_1 \to E_0 [-i, 1, 0] e^{ikz}$, $E_1 \to E_0[1, i, 0]e^{ikz}$. This is the textbook circularly polarized plane wave (of positive helicity), in which the electric and magnetic fields and the propagation direction are mutually perpendicular, and the electric and magnetic fields are in phase quadrature. Theorem 4.3 of Appendix 4A shows that this is only possible in the plane wave limit: transversely finite beams which are everywhere circularly polarized in a fixed plane do not exist.

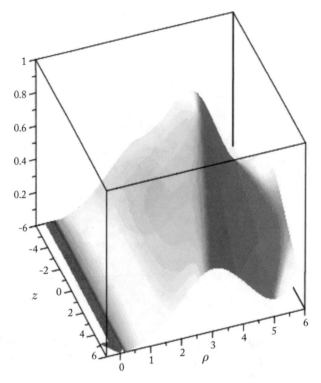

Figure 4.10: The polarization measure Λ for the $m = 1$ TM+iTE beam based on ψ_1. $\Lambda = 0$ (corresponding to circular polarization) on the beam axis $\rho = 0$, as shown in the text. $\Lambda = 1$ at the top of the prominent peak, corresponding to a ring centered on the beam axis on which the electric and magnetic fields are linearly polarized (compare Figure 4.9). The arrow gives the direction of beam propagation.

The beam derived from the vector potential A_1 has B_1 with different polarization properties to that of E_1. Also the electromagnetic energy and momentum densities defined in (3.8) oscillate in time, at angular frequency $2\omega = 2ck$. Self-dual (or 'steady') beams, introduced in Section 3.6, in which the complex fields satisfy $E = \pm i\, B$, may be constructed from the above vector potential. We take

$$A = \frac{1}{2}(A_1 + k^{-1}\nabla \times A_1) = \frac{1}{2}k^{-2}E_0\left[-(\partial_z + ik),\, -i\,(\partial_z + ik),\, \partial_x + i\partial_y\right]\psi$$

$$= \frac{1}{2}k^{-2}E_0 e^{i\phi}\left(-(\partial_z + ik),\, -i\,(\partial_z + ik),\, \partial_\rho - m\rho^{-1}\right)\psi. \tag{4.27}$$

The corresponding magnetic field $\boldsymbol{B} = \nabla \times \boldsymbol{A}$ is

$$
\begin{aligned}
\boldsymbol{B} &= \frac{1}{2} k^{-2} E_0 \left[\partial_y (\partial_x + i\partial_y) + i(\partial_z + ik) \partial_z, \ -\partial_x (\partial_x + i\partial_y) - (\partial_z + ik) \partial_z, \right. \\
& \quad \left. -i(\partial_x + i\partial_y)(\partial_z + ik) \right] \psi \\
&= \frac{1}{2} k^{-2} E_0 e^{i\phi} \left(i(m+1)\rho^{-1}(\partial_\rho - m\rho^{-1}) + i(\partial_z + ik) \partial_z, \right. \\
& \quad \left. -(\partial_\rho^2 - m\rho^{-1}\partial_\rho + m\rho^{-2}) - (\partial_z + ik)\partial_z, \ -i(\partial_\rho - m\rho^{-1})(\partial_z + ik) \right) \psi.
\end{aligned}
\tag{4.28}
$$

The electric field, for waves satisfying the Helmholtz equation, is $\boldsymbol{E} = i\boldsymbol{B}$, as may be verified from the free-space, time-harmonic version of Ampere's law, $\boldsymbol{E} = \frac{i}{k} \nabla \times \boldsymbol{B}$. In the plane wave limit we regain the textbook circularly polarized plane wave, as we had before:

$$
\psi \to e^{ikz}, \qquad \boldsymbol{B} \to E_0 [-i, 1, 0] \, e^{ikz}, \qquad \boldsymbol{E} \to E_0 [1, i, 0] e^{ikz}.
\tag{4.29}
$$

Since $\boldsymbol{E} = i\boldsymbol{B}$ the polarization properties are the same for the electric and magnetic fields. Lekner (2003) [11] has calculated the form taken by Λ for a self-dual 'CP' beam based on a wavefunction, such as ψ_0, which does not depend on the azimuthal angle ϕ. Let suffixes ρ, z denote differentiations. Then

$$
\Lambda =
$$

$$
\frac{\left| \psi_{\rho\rho}^2 - \rho^{-2}\psi_\rho^2 + 2\left(\psi_{\rho\rho} - \rho^{-1}\psi_\rho\right)\left(\psi_{zz} + ik\psi_z\right) - \left(\psi_{\rho z} + ik\psi_\rho\right)^2 \right|}{2|\psi_{zz} + ik\psi_z|^2 + |\psi_{\rho z} + ik\psi_\rho|^2 + 2\mathrm{Re}\left\{ (\psi_{zz} + ik\psi_z)(\psi_{\rho\rho}^* + \rho^{-1}\psi_\rho^*) \right\} + |\psi_{\rho\rho}|^2 + \rho^{-2}|\psi_\rho|^2}.
\tag{4.30}
$$

The limiting fields given in (4.29) give $\boldsymbol{E} \times \boldsymbol{B} = [0, 0, 1]$, and $j_z = (4\pi c)^{-1}$ $[\boldsymbol{r} \times (\boldsymbol{E} \times \boldsymbol{B})]_z = 0$. A finite z component of the angular momentum can result from a transversely finite beam. When the beam is nearly a plane wave but with a small longitudinal component one finds $ck J_z' \approx U'$ (Jackson 1975 Problems 7.20, 7.21 [10]). Hence we can associate a *positive helicity* with finite fields that approach the limit (4.29) within a localized region.

The energy, momentum and angular momentum per unit length of a general 'CP' beam are derived in Appendix 4B.

4.5   SELF-DUAL 'CP' BEAM DERIVED FROM $\psi_0(\rho, z)$

The proto-beam of Lekner 2016 [13] is the confluent form of two classes of beams both with no azimuthal dependence, $m = 0$, as discussed in Section 2.5. The wavefunction is given by

$$
\psi_0(\rho, z) = \frac{2}{k^2} \int_0^k d\kappa \, \kappa \, e^{iqz} J_0(\kappa\rho) = \frac{2}{k^2} \int_0^k dq \, q \, e^{iqz} J_0(\kappa\rho).
\tag{4.31}
$$

From (4.31) the wavenumber weight function is $f_0(k, \kappa) = 2\kappa/k^2$, and from (4B.15) we find, in agreement with Equation 6.16 of Lekner 2016 [13] up to the factor $k^{-2}2\pi/8\pi = k^{-2}/4$, which was omitted there,

$$U' = (17/24)\, k^{-2} E_0^2, \qquad cP'_z = (31/60)\, k^{-2} E_0^2, \qquad ckJ'_z = (31/30)\, k^{-2} E_0^2. \qquad (4.32)$$

Independent confirmation is obtained from spatial integration, as follows. The electric and magnetic fields depend on ψ_0 and on its first and second derivatives with respect to both ρ and z. Expanding $\psi_0(\rho, z)$ in powers of z, we can evaluate all the coefficients of z^n in terms of the regular cylindrical and spherical Bessel functions; to have all the derivatives needed in the focal plane $z = 0$ it is sufficient to go to second order. For example, the components of the complex magnetic field in the focal plane are

$$B_\rho/E_0 e^{i\phi} = -i\left[(k\rho)^{-1} j_1(k\rho) + 3(k\rho)^{-2} J_2(k\rho)\right]$$
$$B_\phi/E_0 e^{i\phi} = (k\rho)^{-1} j_1(k\rho) + (k\rho)^{-2} [k\rho J_1(k\rho) - J_2(k\rho)] \qquad (4.33)$$
$$B_z/E_0 e^{i\phi} = -(k\rho)^{-1} j_2(k\rho) - (k\rho)^{-1} J_2(k\rho).$$

We note that B_ρ is in phase quadrature to the other components. The electric field components are found from $\boldsymbol{E} = i\boldsymbol{B}$. The magnetic field components given in (4.33) are plotted in Figure 4.11.

The measure of linear polarization $\Lambda(\boldsymbol{r}) = |\boldsymbol{B} \cdot \boldsymbol{B}|/\boldsymbol{B} \cdot \boldsymbol{B}^*$ given in (4.30) is the same for the electric and magnetic fields. In the focal plane it is, because of the phase quadrature of B_ρ to the other components, $\Lambda = 1 - 2|B_\rho|^2/|\boldsymbol{B}|^2$, unity when $B_\rho = 0$. From (4.33) we find that this is at $k\rho \approx 4.827, 8.040, \ldots$. Thus circles of exactly linear polarization surround the central circularly polarized region; seen in Lekner 2016 [13] and Figure 4.12 below. On these $\Lambda(\boldsymbol{r}) = 1$ circles the polarization direction is that of $(0, B_\phi, B_z)$ evaluated at the appropriate zero of B_ρ.

The energy and momentum densities are found from (3.60). The integrals needed in the evaluation of the energy, momentum and angular momentum per unit length of the beam are all special cases of the critical form of the Weber-Schafheitlin integral (Watson (1944), Section 13.41 [20]),

$$\int_0^\infty dX\, \frac{J_\mu(X)\, J_\nu(X)}{X^\lambda} = \frac{\Gamma(\lambda)\Gamma\left(\dfrac{\mu + \nu - \lambda + 1}{2}\right)}{2^\lambda \Gamma\left(\dfrac{\lambda + \mu + \nu + 1}{2}\right)\Gamma\left(\dfrac{\lambda - \mu + \nu + 1}{2}\right)\Gamma\left(\dfrac{\lambda + \mu - \nu + 1}{2}\right)},$$
$$Re\,(\mu + \nu + 1) > Re\,(\lambda) > 0.$$

$$(4.34)$$

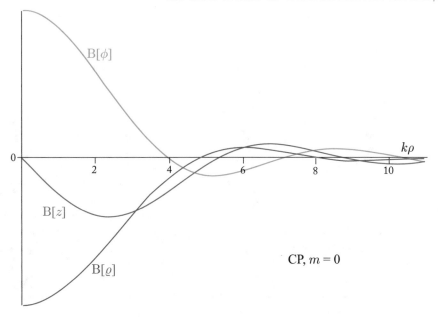

Figure 4.11: Magnetic field components of the 'CP' beam based on ψ_0, in the focal plane $z = 0$. The common phase factor $e^{i\phi}$ is omitted, as is the factor i of B_ρ.

With the use of $j_n(z) = \sqrt{\pi/2z}\, J_{n+\frac{1}{2}}(z)$, $z\Gamma(z) = \Gamma(z+1)$, $\Gamma(1) = 1$, $\Gamma(\frac{1}{2}) = \sqrt{\pi}$ we find, for example,

$$\int_0^\infty dX\, \frac{J_1(X)^2}{X} = \frac{1}{2}, \quad \int_0^\infty dX\, \frac{J_2(X)^2}{X} = \frac{1}{4}, \quad \int_0^\infty dX\, \frac{J_2(X)^2}{X^3} = \frac{1}{24}$$

$$\int_0^\infty dX\, j_1(X)\, J_2(X) = \frac{2}{3}, \quad \int_0^\infty dX\, \frac{j_0(X)\, J_2(X)}{X} = \frac{1}{6}, \quad \int_0^\infty dX\, \frac{j_1(X)\, J_2(X)}{X^2} = \frac{1}{10}$$

$$\int_0^\infty dX\, j_0(X)\, j_1(X) = \frac{1}{2}, \quad \int_0^\infty dX\, \frac{j_1(X)^2}{X} = \frac{1}{4}, \quad \int_0^\infty dX\, \frac{j_2(X)^2}{X} = \frac{1}{12}.$$

$$(4.35)$$

The Weber–Schafheitlin integral enables evaluation of the energy, momentum and angular momentum per unit length of the beam. The results agree with those in (4.32) found by integrating over the wavenumber weight function.

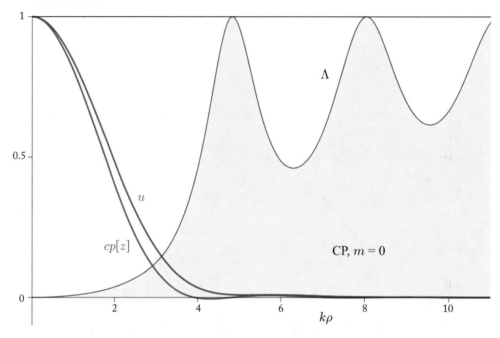

Figure 4.12: Energy density u, z component of the momentum density p_z, and the measure of linear polarization Λ, in the focal plane of the 'CP' beam based on ψ_0. Λ is zero (the beam is circularly polarized) at the center of the focal region, but rises to unity on circles on which the beam is linearly polarized. The first two of these are at $k\rho \approx 4.827,\ 8.040$.

4.6 SELF-DUAL 'CP' BEAM DERIVED FROM $\psi_1(\rho, \phi, z)$

Next we look at the 'CP' beam derived from an $m = 1$ wave function ψ_1. As in Section 2.8, we shall take the wavefunction

$$\psi_1 = -k^{-1}e^{i\phi}\partial_\rho\psi_0 = e^{i\phi}\int_0^k d\kappa\, f_1(k,\kappa)\, e^{iqz} J_1(\kappa\rho). \qquad (4.36)$$

The weight function of ψ_1 is

$$f_1(k,\kappa) = \kappa k^{-1} f_0(k,\kappa) = 2\kappa^2/k^3. \qquad (4.37)$$

With the weight function as defined in (4.37), from (4B.15) we find the energy, momentum and angular momentum per unit length of the beam to be

$$U' = (3/10)\, k^{-2} E_0^2, \qquad c P_z' = (5/28)\, k^{-2} E_0^2, \qquad ck J_z' = (117/140)\, k^{-2} E_0^2. \qquad (4.38)$$

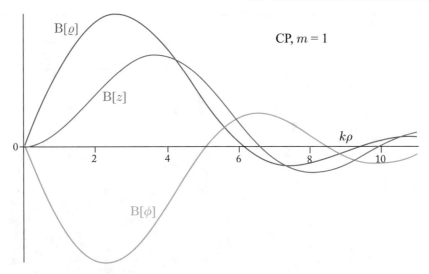

Figure 4.13: Magnetic field components of the 'CP' beam based on ψ_1, in the focal plane $z = 0$. The common phase factor $e^{2i\phi}$ is omitted, as is the factor i of B_ρ.

The physical electromagnetic 'CP' beam derived from ψ_1 has fields given by (4.28). These are, in the focal plane,

$$B_\rho/E_0 e^{2i\phi} = i\left[(k\rho)^{-1} j_2(k\rho) - 4(k\rho)^{-3}\{k\rho J_1(k\rho) - 4J_2(k\rho)\}\right]$$
$$B_\phi/E_0 e^{2i\phi} = -(k\rho)^{-1} j_2(k\rho) + (k\rho)^{-4}\left\{k\rho\left[(k\rho)^2 - 8\right]J_0(k\rho) - 4\left[(k\rho)^2 - 4\right]J_1(k\rho)\right\}$$
$$B_z/E_0 e^{2i\phi} = (k\rho)^{-1} j_3(k\rho) + (k\rho)^{-1}J_3(k\rho). \tag{4.39}$$

The energy, momentum and angular momentum content per unit length of the beam may be evaluated as spatial integrals by using (4B.1), (4.39) and the Weber–Schafheitlin integral (4.34). The results are in agreement with (4.38). The fields given in (4.39) are plotted in Figure 4.13.

We see from (4.39) that again B_ρ is in phase quadrature to the other components. In the focal plane the measure of linear polarization $\Lambda(r)$ is unity when $B_\rho = 0$, namely at $k\rho \approx 6.11$, 9.43, …. As for the $m = 0$ beam, circles of exactly linear polarization surround the central circularly polarized region, as is shown in Figure 4.14. On these $\Lambda(r) = 1$ circles the polarization direction is that of $(0, B_\phi, B_z)$ evaluated at the appropriate zero of B_ρ.

4.7   'LINEARLY POLARIZED' BEAMS

The TM and TE beams of Section 4.2 have exact linear polarization in the magnetic and electric fields, respectively. The respective field lines are circles concentric with the beam axis: the direction of polarization varies continuously with the azimuthal angle. Further, the beam vanishes in

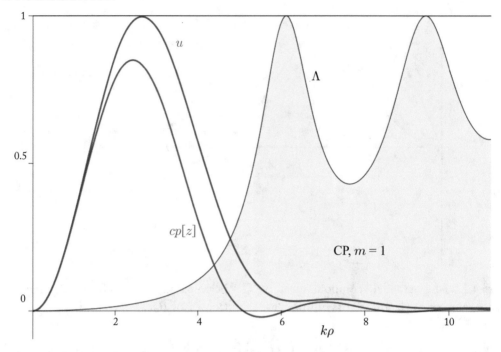

Figure 4.14: Energy density u, z component of the momentum density p_z, and the measure of linear polarization Λ, in the focal plane of the 'CP' beam based on ψ_1. The energy and momentum densities are zero on the beam axis $\rho = 0$. Λ is zero (the beam is circularly polarized) at the center of the focal region, but rises to unity on circles on which the beam is linearly polarized. The smallest of these is at $k\rho \approx 6.11$, as described in the text. The value of the linear polarization measure at the maximum of u (at $k\rho \approx 2.646$) is $\Lambda \approx 0.0286$.

the plane-wave limit: setting $\psi = e^{ikz}$ gives identically zero fields. We wish to construct beams which are linearly polarized in a fixed plane, at least in the plane-wave limit. (Theorem 4.2 of Appendix 4A shows that finite beams which are everywhere linearly polarized in a fixed plane do not exist.)

We shall use both Cartesian $[x, y, z]$ and cylindrical (ρ, ϕ, z) coordinate systems. The vector potential complex amplitude

$$\boldsymbol{A} = A_0 [1, 0, 0] \psi = A_0(\cos\phi, -\sin\phi, 0)\psi \qquad (4.40)$$

gives the complex field amplitudes in Cartesian coordinates,

$$\boldsymbol{B} = \nabla \times \boldsymbol{A} = A_0 \left[0, \partial_z, -\partial_y\right] \psi. \qquad (4.41)$$

$$\boldsymbol{E} = ik^{-1}\nabla(\nabla \cdot \boldsymbol{A}) + ik\boldsymbol{A} = ik^{-1}A_0[\partial_x^2 + k^2, \partial_x\partial_y, \partial_x\partial_z]\psi. \qquad (4.42)$$

In the plane-wave limit $\psi \to e^{ikz}$ these become $\boldsymbol{B} \to ikA_0 [0, 1, 0] e^{ikz}, \boldsymbol{E} \to ikA_0$ $[1, 0, 0] e^{ikz}$. This is the textbook linearly polarized beam, with both fields linearly polarized everywhere and mutually perpendicular. As noted above, these properties cannot persist in transversely localized beams. In the following we explore the properties of transversely finite beams with beam wavefunctions having the azimuthal dependence $e^{im\phi}$, as in (3.13). The complex magnetic field amplitude is then, in cylindrical coordinates, and omitting the factor A_0,

$$\boldsymbol{B} = (\sin\phi\, \partial_z,\ \cos\phi\, \partial_z,\ -\sin\phi\, \partial_\rho - im\rho^{-1}\cos\phi\ )\psi. \tag{4.43}$$

The electric field amplitude is more complicated, so we shall give the components individually:

$$kE_\rho = i\cos\phi\ \left(\partial_\rho^2 + k^2\right)\psi + m\sin\phi\ \left(\rho^{-1}\partial_\rho - \rho^{-2}\right)\psi$$
$$kE_\phi = m\cos\phi\ \left(-\rho^{-1}\partial_\rho + \rho^{-2}\right)\psi + i\sin\phi\ \left(-\rho^{-1}\partial_\rho + m^2\rho^{-2} - k^2\right)\psi \tag{4.44}$$
$$kE_z = i\cos\phi\ \partial_\rho\partial_z\psi + m\sin\phi\ \rho^{-1}\partial_z\psi.$$

We saw above that the plane-wave limit gives linear polarization of the electric field, along the x axis. Let us set $\phi = \pi/2$ or $3\pi/2$ in the field amplitudes. This gives

$$\boldsymbol{B}\,(\cos\phi\ = 0) = (\partial_z,\ 0,\ -\partial_\rho)\psi$$
$$\boldsymbol{E}\,(\cos\phi\ = 0) = \left(m\left(\rho^{-1}\partial_\rho - \rho^{-2}\right),\ i\left(-\rho^{-1}\partial_\rho + m^2\rho^{-2} - k^2\right),\ m\rho^{-1}\partial_z\right)\psi. \tag{4.45}$$

Hence when $\cos\phi\ = 0$ and with $m = 0$ there is only one non-zero electric field component, the azimuthal one. This corresponds to the cartesian component E_x: in the $x = 0$ plane the $m = 0$ 'LP' beams are exactly linearly polarized, along x. For non-zero m the polarization is more complicated, but a strong degree of linear polarization persists in the neighborhood of the beam axis, as we shall see in Section 4.9.

In this paragraph we specialize to $m = 0$ 'LP' beams. The wavefunction depends only on ρ and z. We denote the derivatives by subscripts, so $\psi_z = \partial_z\psi$, et cetera. The field amplitudes in Cartesian coordinates are then

$$\boldsymbol{B} = \left[0,\ \psi_z,\ -\sin\phi\ \psi_\rho\right]$$
$$-ikE_x = \cos^2\phi\ \psi_{\rho\rho} + \rho^{-1}\sin^2\phi\ \psi_\rho + k^2\psi$$
$$-ikE_y = \sin\phi\cos\phi\ (\psi_{\rho\rho} - \rho^{-1}\psi_\rho) \tag{4.46}$$
$$-ikE_z = \cos\phi\ \psi_{\rho z}.$$

In $m = 0$ 'LP' beams the magnetic field is linearly polarized along $\hat{\boldsymbol{y}}$ in the $y = 0$ plane ($\sin\phi = 0$), while the electric field is linearly polarized along $\hat{\boldsymbol{x}}$ in the $x = 0$ plane ($\cos\phi = 0$). A general expression for the polarization measure Λ of $m = 0$ 'LP' beams was given in Lekner 2003, Equation (27) [11].

The cycle-averaged energy and momentum densities may be calculated from the complex field amplitudes using (3.10) and (3.11). In the evaluation of U', cP'_z and cJ'_z we need first to integrate over ϕ in, for example, $U' = \int d^2 r\, \bar{u} = \int_0^\infty d\rho\, \rho \int_0^{2\pi} d\phi\, \bar{u}$. After that the methods of Appendix 4B give us

$$
\begin{bmatrix} U' \\ cP'_z \\ cJ'_z \end{bmatrix} = \frac{E_0^2}{8k^3} \int_0^k d\kappa\, |f(k,\kappa)|^2 \left(k^2 + q^2\right) \kappa^{-1} \begin{bmatrix} k \\ q \\ m \end{bmatrix} \qquad \text{('LP')}. \qquad (4.47)
$$

For the 'LP' beam discussed in the next Section, based on the $m = 0$ proto-beam wavefunction, $f(k,\kappa) = 2\kappa k^{-2}$, and $(U',\, cP'_z) = k^{-2} E_0^2 (3/8,\, 4/15)$, with $cP'_z/U' = 32/45 \approx 0.71$.

4.8   'LINEARLY POLARIZED' BEAM DERIVED FROM ψ_0

We shall give the fields and polarization properties of the 'LP' beam based on the wavefunction ψ_0 given in (4.31) and the vector potential (4.40), in which we omit the common factor A_0. We also omit a common factor 2 in the fields.

We consider the focal plane properties first. The complex magnetic amplitude has the Cartesian components

$$
\begin{aligned}
B_x &= 0, \qquad B_y = 2ixy\,(k\rho)^{-5}\,(\sin k\rho - k\rho\,\cos k\rho) = 2ixy(k\rho)^{-3} j_1(k\rho) \\
B_z &= y(k\rho)^{-3}\,[2J_1(k\rho) - k\rho J_0(k\rho)] = y(k\rho)^{-2} J_2(k\rho).
\end{aligned} \qquad (4.48)
$$

The magnetic field in the focal plane is zero on $y = 0$, since all the components are zero. The magnetic field is linearly polarized along the $\hat{z}$ direction on $x = 0$, and also on the zeros of $\sin k\rho - k\rho\,\cos k\rho$. It is linearly polarized along the $\hat{y}$ direction on the zeros of $J_2(k\rho)$. There are curves of circular polarization when the magnitudes of B_y and B_z (which are in phase quadrature) are equal.

The complex electric field Cartesian components are, in the focal plane,

$$
\begin{aligned}
E_x &= i\,(k\rho)^{-5} \left\{ \left[(ky)^4 + (ky)^2[(kx)^2 - 2] + 6(kx)^2 \right] J_1(k\rho) - k\rho \left[3(kx)^2 - (ky)^2 \right] J_0(k\rho) \right\} \\
E_y &= -ik^2 xy\,(k\rho)^{-5} \left\{ \left[(k\rho)^2 - 8 \right] J_1(k\rho) + 4k\rho J_0(k\rho) \right\} = ik^2 xy(k\rho)^{-3} J_3(k\rho) \\
E_z &= x\,(k\rho)^{-5} \left\{ [3 - (k\rho)^2]\sin k\rho - 3k\rho\,\cos k\rho \right\} = x\,(k\rho)^{-3} j_2(k\rho).
\end{aligned} \qquad (4.49)
$$

The real fields have the time-dependence $\sin \omega t$ for E_x and E_y and $\cos \omega t$ for E_z. Hence the electric field vector in the focal plane cycles between being purely transverse and purely longitudinal. (In the $x = 0$ plane the field is always transverse, as discussed above in Section 4.7.) Figure 4.15 shows the electric field in its transverse mode, together with the cycle-averaged energy density contours.

The electric field is linearly polarized along $\hat{x}$ in the $x = 0$ plane, since both the other components are then zero. However, when $y = 0$ only E_x and E_z are non-zero, and these fields are in phase quadrature. Hence when the E_x and E_z components have equal magnitude the

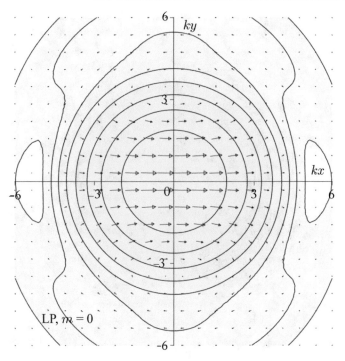

Figure 4.15: The cycle-averaged energy density (contours) and the transverse electric field (arrows), in the focal plane of the 'LP' proto-beam. The logarithm of the energy density is plotted, to show the local minima (not zeros) at $k\,|x| \approx 5.29$, $y = 0$.

electric field is exactly circularly polarized. This happens at $k\,|x| \approx 3.98$, $5.44, \ldots.$. Figure 4.16 shows the complex pattern of the polarization measure Λ. A larger view of the same beam polarization is given in Figure 5 of Andrejic and Lekner 2017, which relates the polarization properties to the phase singularities, and compares the polarizations of the 'LP' beams based on the proto-beam, and on the Carter beam and ψ_b of Section 2.5.

4.9   'LINEARLY POLARIZED' BEAM DERIVED FROM ψ_1

We briefly discuss the properties of the 'LP' beam based on the wavefunction (4.36). The electric field components for general ψ are given in (4.44). As noted in Section 4.7, for non-zero m the polarization is more complicated, but a strong degree of linear polarization persists. The electric field is no longer predominantly along x. Figure 4.17 shows the focal plane energy density and the transverse electric field.

In the focal plane along $y = 0$ (and thus with $\rho = |x|$) the complex electric amplitude is of the form $E_\rho = 3i\,F(k\rho)$, $E_\phi = F(k\rho)$, $E_z = G(k\rho)$. The functions F and G are given by

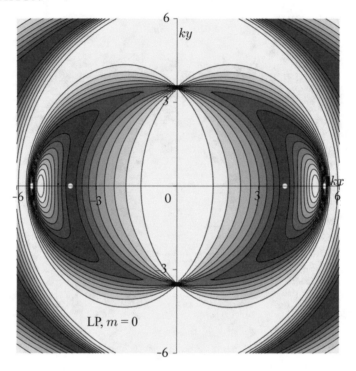

Figure 4.16: The polarization measure Λ_E, in the focal plane of the 'LP' proto-beam. The yellow corresponds to Λ_E near unity (linear polarization), brown to Λ_E near zero (circular polarization). Note that the central focal region of this 'linearly polarized' beam has linear polarization, but that there are localized regions of nearly circular polarization. The electric field is exactly circularly polarized at curves which intersect the focal plane at the points $k\,|x| \approx 3.98,\; 5.44$. There is an in-plane circle of perfect linear polarization at $k\rho \approx 5.76$, originating in a zero of $j_2(k\rho)$ and the consequent zero of E_z, which leaves the in-phase transverse components. Along $y = 0$, lines of perfect linear polarization intersect the focal plane at $k\,|x| \approx 5.14$, originating in a zero of E_x, which leaves the longitudinal component.

(we omit a constant common factor)

$$
\begin{aligned}
F &= (k\rho)^{-4}\left\{4k\rho J_0\,(k\rho) + \left[(k\rho)^2 - 8\right]J_1(k\rho)\right\} = -(k\rho)^{-2}J_3(k\rho)\\
G &= (k\rho)^{-5}\left\{k\rho\left[12 - (k\rho)^2\right]\cos k\rho + \left[5(k\rho)^2 - 12\right]\sin k\rho\right\}\\
&= (k\rho)^{-2}\left[j_2(k\rho) - k\rho j_3(k\rho)\right].
\end{aligned}
\tag{4.50}
$$

When $F\,(k\rho) = 0$ the electric field is linearly polarized along z. $F\,(0) = 0$ and hence the beam is longitudinal and linearly polarized on the beam axis. Also discernible in Figure 4.17

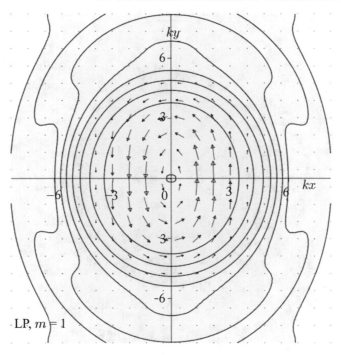

Figure 4.17: The cycle-averaged energy density (contours) and the transverse electric field (arrows), in the focal plane of the 'LP' beam based on ψ_1. The logarithm of the energy density is plotted.

is the next root of F, at $k\,|x| \approx 6.38$. When $8F^2 = G^2$ the imaginary radial component has the same magnitude as the vector sum of the real azimuthal and longitudinal components. The electric field is then circularly polarized. This happens at $k\rho \approx 0.969$, 5.03, and 6.78, visible as dark shading in Figure 4.18. The perfect circular polarization impinges as close as 0.154 of the vacuum wavelength to the beam axis.

In the focal plane along $x = 0$ (and thus with $\rho = |y|$) the complex electric amplitude is of the form $E_\rho = -iF(k\rho)$, $E_\phi = f(k\rho)$, $E_z = g(k\rho)$. F is defined above; the functions f and g are given by (we again omit a constant common factor)

$$
\begin{aligned}
f &= (k\rho)^{-4} \left\{ k\rho[(k\rho)^2 - 4]J_0\,(k\rho) + \left[8 - 3(k\rho)^2\right] J_1(k\rho) \right\} \\
g &= (k\rho)^{-5} \left\{ \left[3 - (k\rho)^2\right] \sin k\rho - 3\cos k\rho \right\} = (k\rho)^{-2} j_2(k\rho).
\end{aligned}
\tag{4.51}
$$

When $f^2 + g^2 = F^2$, as happens at $k\,|y| \approx 4.65$, $5.12\ldots$, the imaginary radial component has the same magnitude as the vector sum of the real azimuthal and longitudinal components. The electric field is then circularly polarized. When $F\,(k\rho) = 0$ the electric field is linearly polarized, because the remaining azimuthal and longitudinal components of the complex amplitude are

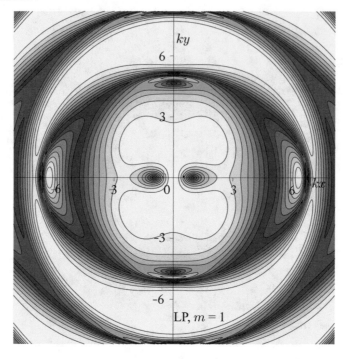

Figure 4.18: The polarization measure Λ_E, in the focal plane of the 'LP' ψ_1 beam. The yellow corresponds to Λ_E near unity (linear polarization), brown to Λ_E near zero (circular polarization). Note that the central focal region of this 'linearly polarized' beam has linear polarization, but that localized regions of nearly circular polarization intrude close to the focal center. The electric field is exactly circularly polarized on curves which intersect the focal plane, the closest intersections being at $k\,|x| \approx 0.969$.

real. This happens at $k\rho = 0$ and at $k\rho \approx 6.38$, as above on the x axis. The polarization is nearly (but not exactly) linear in the neighborhood of the circle $k\rho \approx 7.1$.

Sections 4.7 to 4.9 are based on Lekner 2020 [14], where more detail about 'linearly polarized' beams may be found.

4A   APPENDIX: NON-EXISTENCE THEOREMS

In textbooks a light beam is usually represented by a plane wave, with $\boldsymbol{E}$, $\boldsymbol{B}$ and the propagation vector $\boldsymbol{k}$ everywhere mutually perpendicular. This 'beam' can be everywhere linearly polarized in the same direction, or everywhere circularly polarized in the same plane, and its energy is everywhere transported in a fixed direction at the speed of light. It has been shown in Lekner

2003 [11] that *none* of these properties can hold for a transversely finite beam. The non-existence theorems are stated and proved below.

Theorem 4.1 *TEM beams do not exist.*

By TEM beams we mean those in which $\boldsymbol{E}$ and $\boldsymbol{B}$ are both transverse to the propagation direction. Suppose that the complex fields are transverse everywhere: $\boldsymbol{E} = [E_x, E_y, 0]$, $\boldsymbol{B} = [B_x, B_y, 0]$. With time dependence e^{-ikct} the Maxwell curl equations in free space become $\nabla \times \boldsymbol{E} = ik\boldsymbol{B}$, $\nabla \times \boldsymbol{B} = -ik\boldsymbol{E}$. The first of these reads

$$[-\partial_z E_y, \partial_z E_x, \partial_x E_y - \partial_y E_x] = ik[B_x, B_y, 0]. \tag{4A.1}$$

Hence $\partial_x E_y - \partial_y E_x = 0$. Taking the curl of (4A.1) gives

$$[-\partial_z^2 E_x, -\partial_z^2 E_x, \partial_z(\partial_x E_x + \partial_y E_y)] = k^2[E_x, E_y, 0]. \tag{4A.2}$$

Also the divergence of $\boldsymbol{E}$ is zero in free space, so $\partial_x E_x + \partial_y E_y = 0$. Hence the non-zero components of $\boldsymbol{E}$ satisfy the equations

$$\partial_z^2 E_x + k^2 E_x = 0, \qquad \partial_z^2 E_y + k^2 E_y = 0. \tag{4A.3}$$

Propagating solutions of (4A.3) have the form $E_x = e^{ikz} F(x, y)$, $E_y = e^{ikz} G(x, y)$. Then $\partial_x E_y - \partial_y E_x = 0$ implies $\partial_x G - \partial_y F = 0$, and $\partial_x E_x + \partial_y E_y = 0$ implies $\partial_x F + \partial_y G = 0$. Thus, F, G are a Cauchy-Riemann pair: they are harmonic functions satisfying $(\partial_x^2 + \partial_y^2) F, G = 0$. Harmonic functions do not have maxima or minima except at the border of their domain. Hence a TEM beam cannot be localized around the beam axis (here the z axis). The statement holds in free space; in waveguides TEM modes can exist in the presence of cylindrical conductors.

Theorem 4.2 *Beams of fixed linear polarization do not exist.*

Suppose $\boldsymbol{E} = [F(x, y, z), 0, 0]$, so the beam is linearly polarized along $\hat{\boldsymbol{x}}$, everywhere. Then from the Maxwell curl equations, with e^{-ikct} time dependence, we have $ik\boldsymbol{B} = \nabla \times \boldsymbol{E} = [0, \partial_z, -\partial_y]F$, and $ik\nabla \times \boldsymbol{B} = [-(\partial_y^2 + \partial_z^2), \partial_x\partial_y, \partial_x\partial_z]F = k^2\boldsymbol{E}$. Hence

$$(\partial_y^2 + \partial_z^2 + k^2) F = 0 \quad \text{and} \quad \partial_x\partial_y F = 0 = \partial_x\partial_z F. \tag{4A.4}$$

The last two equations imply $F(x, y, z) = f(x) + g(z) + h(y, z)$, which cannot represent a beam localized transversely in the x direction. Thus, a localized beam transversely polarized in the same direction everywhere cannot exist.

If instead we choose longitudinal polarization, $\boldsymbol{E} = [0, 0, F(x, y, z)]$, we find

$$(\partial_x^2 + \partial_y^2 + k^2) F = 0 \quad \text{and} \quad \partial_x\partial_z F = 0 = \partial_y\partial_z F. \tag{4A.5}$$

The last two equations imply $F(x, y, z) = f(y) + g(z) + h(x, y)$, in which $h(x, y)$ may be localized transversely, but cannot represent a beam propagating along the z direction.

Theorem 4.3   *Beams which are everywhere circularly polarized in the same plane do not exist.*

The conditions of circular polarization are that the real and imaginary parts of the field amplitude be perpendicular, and of equal magnitude. Suppose E to lie in the xy plane, and take $E_r = [F(r), 0, 0]$, $E_i = [0, F(r), 0]$. The Maxwell curl equations give us, for monochromatic waves,

$$\nabla \times E_r + k B_i = 0, \qquad \nabla \times E_i - k B_r = 0. \qquad (4A.6)$$
$$\nabla \times B_r - k E_i = 0, \qquad \nabla \times B_i + k E_r = 0. \qquad (4A.7)$$

Hence $k B_r = [-\partial_z F, 0, \partial_x F]$, $k B_i = [0, -\partial_z F, \partial_y F]$. Substitution back into the curl equations (4A.6) then gives

$$\left(\partial_x^2 + \partial_z^2 + k^2\right) F = 0, \qquad \partial_x \partial_y F = 0 = \partial_y \partial_z F. \qquad (4A.8)$$

Hence $F = F(x, z)$. Like wise the Equations (4A.7) give

$$\left(\partial_y^2 + \partial_z^2 + k^2\right) F = 0, \qquad \partial_x \partial_y F = 0 = \partial_x \partial_z F. \qquad (4A.9)$$

These imply $F = F(y, z)$. Thus, F must be a function of z only: only plane waves can be circularly polarized everywhere in the same plane. Note that the beam propagation direction was not specified: the proof applies to circular polarization in any fixed plane.

Theorem 4.4   *Beams or pulses within which the energy velocity is everywhere in the same direction and of magnitude c do not exist.*

We need to define 'energy velocity' first. In fluid flow the conservation of matter requires that $\partial_t \rho + \nabla \cdot (\rho v) = 0$. Here ρ is the fluid density, and v the fluid velocity. In electrodynamics the analogous conservation of energy relation reads $\partial_t u + \nabla \cdot S = 0$, where $u = (8\pi)^{-1}(E^2 + B^2)$ is the energy density, and the Poynting vector $S = (c/4\pi)E \times B$ is the energy flux density (E, B are real fields in this section). A possible definition of the velocity at which energy is transferred is thus (Lekner 2003 [11])

$$v_e = \frac{S}{u} = 2c\frac{E \times B}{E^2 + B^2}. \qquad (4A.10)$$

Note that the magnitude of v_e is necessarily less than or equal to c:

$$\frac{v_e^2}{c^2} = 4\frac{E^2 B^2 - (E \cdot B)^2}{(E^2 + B^2)^2}, \qquad 1 - \frac{v_e^2}{c^2} = \frac{\left(E^2 - B^2\right)^2 + 4(E \cdot B)^2}{(E^2 + B^2)^2} \geq 0. \qquad (4A.11)$$

Suppose, contrary to the statement of Theorem 4.4, that $v_e = c$ everywhere. Then $E^2 = B^2$ and $E \cdot B = 0$, everywhere. Let $E \times B$ point in the z direction. Then $E_y B_z = B_y E_z$, $E_z B_x = B_z E_x$ and $E_x B_y - B_x E_y = (E^2 + B^2)/2$. These three conditions, together with $E^2 = B^2$ and $E \cdot B = 0$ have only one real solution set, namely

$$E_x = B_y, \qquad E_y = -B_x, \qquad E_z = 0 = B_z. \tag{4A.12}$$

This solution set is consistent with Maxwell's equations provided that

$$(\partial_z + \partial_{ct}) B_x = 0, \qquad (\partial_z + \partial_{ct}) B_y = 0 \tag{4A.13}$$

$$\partial_x B_y - \partial_y B_x = 0, \qquad \partial_x B_x + \partial_y B_y = 0. \tag{4A.14}$$

The first two are satisfied by B_x, B_y being arbitrary functions of $z - ct$. But the second pair of equations imply that B_x and B_y are harmonic functions (compare Theorem 4.1), and thus cannot be localized in the x or y directions (these are transverse to the propagation direction, by assumption). Only plane wave pulses or beams can have v_e in the same direction and of magnitude c everywhere.

4B   APPENDIX: ENERGY, MOMENTUM, AND ANGULAR MOMENTUM PER UNIT LENGTH OF 'CP' BEAMS

The energy, momentum and angular momentum per unit length are obtained by integrating over a section of the beam at constant z $\left(\int d^2 r = \iint_{-\infty}^{\infty} dx dy = \int_0^{\infty} d\rho\, \rho \int_0^{2\pi} d\phi \right)$:

$$U' = \int d^2 r\, u(r), \quad P' = \int d^2 r\, p(r), \quad J' = \int d^2 r\, r \times p(r). \tag{4B.1}$$

We now consider 'CP' beams when the azimuthal dependence of ψ is $e^{im\phi}$. The energy, momentum and angular momentum densities can be calculated from (4B.1), using the magnetic field given in (4.28). They are too long to be usefully written down, but are easily generated by computer algebra. There is ρ dependence but no ϕ dependence in the integrands. We use the recurrence formulae (Watson 1944, Section 2.12 [20])

$$2 J'_m(\zeta) = J_{m-1}(\zeta) - J_{m+1}(\zeta), \qquad \frac{2m}{\zeta} J_m(\zeta) = J_{m-1}(\zeta) + J_{m+1}(\zeta). \tag{4B.2}$$

After the elimination of derivatives, the energy and momentum densities contain products of Bessel functions and inverse powers of ρ, of the forms

$$\begin{aligned}
T_0 &= J_m(\kappa\rho) J_m(\kappa'\rho), \\
T_1 &= (m+1)\rho^{-1} \left\{ \kappa' J_{m+1}(\kappa\rho) J_m(\kappa'\rho) + \kappa J_m(\kappa\rho) J_{m+1}(\kappa'\rho) \right\}, \\
T_2 &= (m+1)^2 \rho^{-2} J_{m+1}(\kappa\rho) J_{m+1}(\kappa'\rho).
\end{aligned} \tag{4B.3}$$

Use of the second recurrence formula in (4B.2) gives us the identities

$$T_1 = \kappa\kappa' \left\{ T_0 + \frac{1}{2} \left[J_{m+2}\left(\kappa\rho\right) J_m\left(\kappa'\rho\right) + J_m\left(\kappa\rho\right) J_{m+2}\left(\kappa'\rho\right) \right] \right\} \tag{4B.4}$$

$$T_2 = \frac{\kappa\kappa'}{4} \left\{ T_0 + J_{m+2}\left(\kappa\rho\right) J_{m+2}\left(\kappa'\rho\right) + J_{m+2}\left(\kappa\rho\right) J_m\left(\kappa'\rho\right) + J_m\left(\kappa\rho\right) J_{m+2}\left(\kappa'\rho\right) \right\}. \tag{4B.5}$$

We see that the combination $T_1 - 2T_2$ which occurs in the energy density reduces to

$$T_1 - 2T_2 = \frac{\kappa\kappa'}{2} \left\{ J_m\left(\kappa\rho\right) J_m\left(\kappa'\rho\right) - J_{m+2}\left(\kappa\rho\right) J_{m+2}\left(\kappa'\rho\right) \right\}. \tag{4B.6}$$

Then all of the terms in the energy integral are of the form T_0. Thus, the integration over ρ can be performed using Hankel's inversion formula (Watson 1944, Section 14.4 [20]; Lekner 2004, Appendix A [12]), which may be written as

$$\int_0^\infty d\rho \, \rho \, J_m\left(\kappa\rho\right) \, J_m\left(\kappa'\rho\right) = \kappa^{-1}\delta\left(\kappa - \kappa'\right) \qquad (\kappa, \kappa' > 0). \tag{4B.7}$$

The integration over ρ of products of Bessel functions of the same order thus selects $\kappa' = \kappa$, and the integration over ϕ gives 2π. The energy per unit length of the beam reduces to

$$U' = \frac{E_0^2}{8k^3} \int_0^k d\kappa |f(k,\kappa)|^2 (k+q)^2 \kappa^{-1} k. \tag{4B.8}$$

The same identities enable us to reduce the spatial integrations of the z component of the momentum content per unit length of the beam to

$$cP_z' = \frac{E_0^2}{8k^3} \int_0^k d\kappa |f(k,\kappa)|^2 (k+q)^2 \kappa^{-1} q. \tag{4B.9}$$

The calculation of the z component of the angular momentum content per unit length of the beam is different: we need to integrate over ρ times $j_z = \rho p_\phi$. The integrand of J_z contains $\rho J_1(\kappa\rho) J_1(\kappa'\rho)$, evaluated by the Hankel integral (4B.7), but also the terms $\rho^2 J_m(\kappa\rho) J_{m+1}(\kappa'\rho)$ and $\rho^2 J_{m+1}(\kappa\rho) J_m(\kappa'\rho)$. Symmetric combinations of these singular integrals over ρ also lead to delta functions, as discussed in Appendix 3A.

From (4B.1) and $j_z = x p_y - y p_x = \rho p_\phi$,

$$J_z' = \int d^2r \, \rho \, p_\phi = \int_0^{2\pi} d\phi \int_0^\infty d\rho \, \rho^2 \, p_\phi. \tag{4B.10}$$

We substitute the expression (4.24) for the wavefunction into the azimuthal component of the momentum density, to find that the ρ dependent part of the integrand in (4B.10) consists of two terms,

$$K_1 = (m+1)\kappa\kappa' \left(2k + q + q'\right) \rho J_{m+1}(\kappa\rho) J_{m+1}(\kappa'\rho)$$

$$K_2 = \left(k+q\right)(k+q') \rho^2 \left[\kappa q' J_{m+1}\left(\kappa\rho\right) J_m\left(\kappa'\rho\right) + \kappa' q J_m(\kappa\rho) J_{m+1}(\kappa'\rho) \right]. \tag{4B.11}$$

The integration over ρ of K_1 is effected using (4B.7), and leads to the following contribution to the angular momentum per unit length of the beam:

$$c J_1' = \frac{E_0^2}{8k^3}(2m+2)\int_0^k d\kappa \, |f(k,\kappa)|^2 (k+q)\kappa. \tag{4B.12}$$

The integrals over K_2, on using the results of Appendix 3A, specifically (3A.10), give us

$$c J_2' = \frac{E_0^2}{8k^3}\int_0^k d\kappa \, |f(k,\kappa)|^2 (k+q)^2 \frac{1}{\kappa q}\left[(2m+2)k^2 - (2m+1)\kappa^2\right]. \tag{4B.13}$$

The total is

$$c J_z' = \frac{E_0^2}{8k^3}\int_0^k d\kappa \, |f(k,\kappa)|^2 (k+q)^2 \kappa^{-1}\left[m+1+\frac{\kappa^2}{2kq}\right]. \tag{4B.14}$$

We summarize the results obtained for the 'CP' beams:

$$\begin{bmatrix} U' \\ c P_z' \\ c J_z' \end{bmatrix} = \frac{E_0^2}{8k^3}\int_0^k d\kappa |f(k,\kappa)|^2 (k+q)^2 \kappa^{-1} \begin{bmatrix} k \\ q \\ m+1+\frac{\kappa^2}{2kq} \end{bmatrix}. \tag{4B.15}$$

Incidentally, we have proved that U', P_z', and J_z' are all independent of z, and verified that $U' > c P_z'$, since $k \geq q = \sqrt{k^2 - \kappa^2}$.

4.12   CITED REFERENCES

[1] Andrejic, P. and Lekner J. 2017. Topology of phase and polarization singularities in focal regions, *J. Opt.*, 19:105609:8.

[2] Berry, M. V. 1998. Wave dislocation reactions in non-paraxial Gaussian beams, *Journal of Modern Optics*, 45:1845–1858. DOI: 10.1080/09500349808231706.

[3] Berry, M. V. 2004. Index formulae for singular lines of polarization, *Journal of Optics A: Pure and Applied Optics*, 6:675–678. DOI: 10.1088/1464-4258/6/7/003.

[4] Berry, M. V. and Dennis, M. R. 2001. Polarization singularities in isotropic random vector waves, *Proc. of the Royal Society of London A*, 457:141–155. DOI: 10.1098/rspa.2000.0660.

[5] Carter, W. H. 1973. Anomalies in the field of a Gaussian beam near focus, *Optics Communications*, 64:491–495. DOI: 10.1016/0030-4018(73)90012-6.

[6] Davis, L. W. and Patsakos, G. 1981. TM and TE electromagnetic beams in free space, *Optics Letters*, 6:22–23. DOI: 10.1364/ol.6.000022.

[7] Dennis, M. R. 2002. Polarization singularities in paraxial vector fields: Morphology and statistics, *Optic Communications*, 213:201–221. DOI: 10.1016/s0030-4018(02)02088-6.

[8] Born, M. and Wolf, E. 1999. *Principles of Optics*, 7th ed., Cambridge University Press. 63, 65

[9] Hurwitz, H. 1945. The statistical properties of unpolarised light, *Journal of the Optical Society of America*, 35:525–531. DOI: 10.1364/josa.35.000525. 64

[10] Jackson, J. D. 1975. *Classical Electrodynamics*, 2nd ed., Wiley, New York. DOI: 10.1063/1.3057859. 77

[11] Lekner, J. 2003. Polarization of tightly focused laser beams, *Journal of Optics A: Pure and Applied Optics*, 5:6–14. DOI: 10.1088/1464-4258/5/1/302. 64, 73, 77, 83, 89, 90

[12] Lekner, J. 2004. Invariants of three types of generalized Bessel beams, *Journal of Optics A: Pure and Applied Optics*, 6:837–843. DOI: 10.1088/1464-4258/6/9/004. 92

[13] Lekner, J. 2016. Tight focusing of light beams: A set of exact solutions, *Proc. of the Royal Society A*, 472(20160538):17. DOI: 10.1098/rspa.2016.0538. 73, 77, 78

[14] Lekner, J. 2020. Properties of linearly polarized electromagnetic beams, *Opt. Commun.*, 466:125667:7. 88

[15] Lindell, I. V. 1992. *Methods for Electromagnetic Field Analysis* Section 1.5, Oxford. 65

[16] Nye, J. F. and Berry, M. V. 1974. Dislocations in wave trains, *Proc. of the Royal Society of London A*, 336:165–190. DOI: 10.1098/rspa.1974.0012.

[17] Nye, J. F. and Hajnal, J. V. 1987. The wave structure of monochromatic electromagnetic radiation, *Proc. of the Royal Society A*, 409:212–36. DOI: 10.1098/rspa.1987.0002. 65

[18] Nye, J. F. 1998. Unfolding of higher-order wave dislocations, *Journal of the Optical Society of America*, 15:1132–1138. DOI: 10.1364/josaa.15.001132.

[19] Nye, J. F. 1999. *Natural Focusing and the Fine Structure of Light*, Institute of Physics Publishing, Bristol. 65

[20] Watson, G. N. 1944. *Theory of Bessel Functions*, 2nd ed., Cambridge University Press. 78, 91, 92

C H A P T E R   5

Chirality

5.1 INTRODUCTION

Chirality of molecules and of crystals leads to optical activity or rotatory power, the ability of a medium to rotate the plane of polarization of light. Its discovery dates back to work by Arago in 1811, Biot (five memoirs between 1812 and 1837) and Fresnel's 1822 conjecture that on entering an optically active medium light is split into two beams of opposite circular polarization which travel with different phase velocities. In 1848 Pasteur demonstrated that the rotatory power of a tartrate solution is related to the form the tartrate crystals take: crystals of opposite handedness dissolve to give solutions with opposite rotatory power (Lowry 1964 [7]).

Here we are concerned with the chirality of light itself, and with the special properties of *self-dual* light beams and of their chiral measures. The free-space Maxwell *equations* are unchanged by the duality transformation $E \to B$, $B \to -E$. However, *solutions* of the Maxwell equations are in general changed by the duality transformation into physically different solutions. For example, a transverse electric (TE) beam is changed into a transverse magnetic (TM) beam. In Section 3.6 we introduced self-dual electromagnetic beams (those unchanged by the duality transformation). In Section 3.7 we defined the TM±iTE beams; the notation is a shorthand for the superposition of the fields of a TM beam and of a TE beam, in phase quadrature. In Section 4.4 we defined self-dual 'circularly polarized' beams; their properties were explored in Sections 4.5 and 4.6.

The literature on chiral light is extensive: see the Additional References at the end of this Chapter, and Lekner 2019 [4]. Lipkin (1964) [6] found conserved quantities which have been interpreted as the chiral density and the chiral current. Tang and Cohen (2010, 2011) [8, 9] have shown that the chiral density determines the asymmetry in the rates of excitation between a chiral molecule and its mirror image. The chiral density $\chi(r,t)$ and chiral current $C(r,t)$ are defined for real fields $E(r,t)$, $B(r,t)$ as

$$\chi = E \cdot (\nabla \times E) + B \cdot (\nabla \times B). \tag{5.1}$$

$$C = E \times (\nabla \times B) - B \times (\nabla \times E). \tag{5.2}$$

We note that both χ and C are unchanged by the duality transformation $E \to B$, $B \to -E$.

The quantity $E \cdot (\nabla \times E)$ is sometimes referred to as the local helicity density of the vector field E, and 'helicity' is often used in this context. We prefer *chiral* density and *chiral* current, since helicity in particle physics is the projection of the angular momentum onto the momentum direction.

In the terminology of this Chapter, χ is the sum of the electric and magnetic field chiral densities. It is clear that the chiral density is maximum when the fields are eigenvectors of curl, with the same eigenvalue, $\nabla \times \mathbf{E} = k\mathbf{E}$, $\nabla \times \mathbf{B} = k\mathbf{B}$. Self-dual fields are eigenvectors of curl, as we saw in Section 3.6. We shall see that it follows from (5.1) that *the chiral density of self-dual fields is maximal, and proportional to the energy density*. Likewise it follows from (5.2) that *the chiral current of self-dual fields is maximal, and proportional to the momentum density* (or the Poynting vector).

Bliokh ad Nori (2011) [1] have shown by means of a Fourier representation of the electric and magnetic fields that the ratio of the chiral density to the energy density in monochromatic beams of angular frequency $\omega = ck$ lies between $\pm 8\pi k$ (using our definition of chiral density). We show in Section 5.5 that for self-dual fields the ratio has to be exactly $\pm 8\pi k$. This gives an alternative proof of the chiral density being maximal for self-dual fields.

5.2   PROPERTIES OF THE CHIRALITY MEASURES OF ELECTROMAGNETIC FIELDS

In the defining relations (5.1) and (5.2) we can replace the curls by time derivatives, from Maxwell's free-space curl equations $\nabla \times \mathbf{E} + \partial_{ct}\mathbf{B} = 0$, $\nabla \times \mathbf{B} - \partial_{ct}\mathbf{E} = 0$. This gives

$$\chi = -\mathbf{E} \cdot \partial_{ct}\mathbf{B} + \mathbf{B} \cdot \partial_{ct}\mathbf{E} \tag{5.3}$$

$$\mathbf{C} = \mathbf{E} \times \partial_{ct}\mathbf{E} + \mathbf{B} \times \partial_{ct}\mathbf{B}. \tag{5.4}$$

For monochromatic electromagnetic beams we have (with a similar decomposition for the magnetic field)

$$\mathbf{E}(\mathbf{r},t) = Re\left\{\mathbf{E}(\mathbf{r})e^{-i\omega t}\right\} = Re\left\{(\mathbf{E}_r + i\,\mathbf{E}_i)e^{-i\omega t}\right\} = \mathbf{E}_r \cos \omega t + \mathbf{E}_i \sin \omega t. \tag{5.5}$$

The angular frequency is $\omega = ck$. Carrying out the time differentiations gives

$$\chi(\mathbf{r}) = k\,(\mathbf{B}_r \cdot \mathbf{E}_i - \mathbf{E}_r \cdot \mathbf{B}_i), \qquad \mathbf{C}(\mathbf{r}) = k\,(\mathbf{E}_r \times \mathbf{E}_i + \mathbf{B}_r \times \mathbf{B}_i). \tag{5.6}$$

Thus the chiral density and the chiral current of monochromatic beams are *independent of time*, in contrast to the energy and momentum densities which in general oscillate at angular frequency 2ω.

Another way of writing (5.6) is in terms of the complex field amplitudes $\mathbf{E}(\mathbf{r})$, $\mathbf{B}(\mathbf{r})$:

$$\chi(\mathbf{r}) = k\,Im\left(\mathbf{E} \cdot \mathbf{B}^*\right), \qquad \mathbf{C}(\mathbf{r}) = \frac{ik}{2}\{\mathbf{E} \times \mathbf{E}^* + \mathbf{B} \times \mathbf{B}^*\}. \tag{5.7}$$

From the defining relations (5.1) and (5.2) and the free-space Maxwell equations it follows that the density and current satisfy the conservation law

$$\partial_{ct}\chi + \nabla \cdot \mathbf{C} = 0 \qquad (\nabla \cdot \mathbf{C} = 0 \text{ for monochromatic beams}). \tag{5.8}$$

Equation 5.8 is the analogue of the energy-momentum conservation law $\partial_{ct} u + c\nabla \cdot \boldsymbol{p} = 0$. It follows from the arguments of Section 1.3 that the chirality current component C_z per unit length of a transversely localized monochromatic electromagnetic beam is constant along the beam: applying $\int d^2r = \int_{-\infty}^{\infty} dx \int_{-\infty}^{\infty} dy = \int_0^{\infty} d\rho\, \rho \int_0^{2\pi} d\phi$ to $\nabla \cdot \boldsymbol{C} = \partial_x C_x + \partial_y C_y + \partial_z C_z = 0$ gives

$$\partial_z \int d^2r\, C_z = 0, \quad \text{or} \quad C_z' = \int d^2r\, C_z = \text{constant.} \tag{5.9}$$

5.3   CHIRALITY OF TM AND TE BEAMS

As the simplest example of chirality, consider the TM (transverse magnetic) monochromatic beams of Section 3.2. These have vector amplitude along the beam axis, $\boldsymbol{A}_{TM} = [0, 0, \psi] = (0, 0, \psi)$. We shall later add the prefactor $A_0 = k^{-1} E_0$ to give the fields amplitude proportional to E_0. Square brackets are used for Cartesian coordinates $[x, y, z]$, round brackets for cylindrical (ρ, ϕ, z) coordinates. In general the ϕ dependence of ψ is in the factor $e^{im\phi}$, as in (2.14). The results we shall find for TM beams apply also to TE beams, since the duality transformation $\boldsymbol{E} \to \boldsymbol{B}$, $\boldsymbol{B} \to -\boldsymbol{E}$ leaves χ and $\boldsymbol{C}$ unchanged.

We shall assume at first that the beam wavefunction does not depend on the azimuthal angle, and set $\psi = \psi\,(\rho, z) = \psi_r\,(\rho, z) + i\psi_i(\rho, z)$, $m = 0$ in (2.14). Then the complex magnetic amplitude is $\boldsymbol{B} = \nabla \times \boldsymbol{A} = (0, -\partial_\rho \psi, 0)$, and the magnetic field is everywhere azimuthal and is linearly polarized, since the real and imaginary parts are collinear. The magnetic field lines are circles, with centers on the beam axis. The complex electric amplitude has radial and longitudinal components, $\boldsymbol{E} = ik^{-1}(\partial_\rho \partial_z, 0, \partial_z^2 + k^2)\psi$, and the electric field is elliptically polarized in general. The electric and magnetic fields are everywhere perpendicular. The azimuthal component of the momentum density is zero, and therefore so is the angular momentum. The real and imaginary parts of the complex field amplitudes are

$$\boldsymbol{B}_r = \left(0, -\partial_\rho \psi_r, 0\right), \qquad \boldsymbol{B}_i = (0, -\partial_\rho \psi_i, 0) \tag{5.10}$$

$$\boldsymbol{E}_r = -k^{-1}\left(\partial_\rho \partial_z, 0, \partial_z^2 + k^2\right) \psi_i, \qquad \boldsymbol{E}_i = k^{-1}\left(\partial_\rho \partial_z, 0, \partial_z^2 + k^2\right) \psi_r. \tag{5.11}$$

From (5.6) the chiral density is zero, since the electric and magnetic fields are everywhere perpendicular. The chiral current is, again from (5.6), $\boldsymbol{C} = (0, C_\phi, 0)$, with C_ϕ proportional to

$$[(\partial_z^2 + k^2)\, \psi_r](\partial_\rho \partial_z \psi_i) - [(\partial_z^2 + k^2)\, \psi_i](\partial_\rho \partial_z \psi_r). \tag{5.12}$$

The chiral current of $m = 0$ TM or TE beams is everywhere azimuthal. In the plane-wave limit, $\psi \to e^{ikz}$, the one non-zero chiral component C_ϕ vanishes. In fact the whole beam vanishes in this limit.

Next we give the general formulae for the chiral density and chiral current for TM or TE beams based on wavefunctions with any integer m. The complex field amplitudes for the general

TM beam are given in (3.15) and (3.16). We write the derivatives of the beam wavefunction as subscripts, for example $\partial_\rho \psi = \psi_\rho$. The chiral density is proportional to m:

$$\chi = -\frac{m}{k\rho} \, Im\left\{ \psi \psi_{\rho z}^* + \psi_\rho \psi_z^* \right\}. \tag{5.13}$$

The chiral current density has radial and longitudinal components proportional to m:

$$C_\rho = \frac{m}{k^2 \rho} \, Re\left\{ (\psi_{zz} + k^2 \psi)\psi_z^* \right\}$$

$$C_\phi = \frac{1}{k^2} \, Im\left\{ (\psi_{zz} + k^2 \psi)\psi_{\rho z}^* \right\} \tag{5.14}$$

$$C_z = \frac{m}{k^2 \rho} \, Re\left\{ \psi_{\rho z} \psi_z^* + k^2 \psi_\rho \psi^* \right\}.$$

We saw in the previous Section that $C_z' = \int d^2r\, C_z$ is constant along the length of the beam. By analogy with energy and momentum, we may expect that the chiral content per unit length of the beam is also constant,

$$X' = \int d^2r \, \chi = \int_0^\infty d\rho \, \rho \int_0^{2\pi} d\phi \, \chi = \text{ constant.} \tag{5.15}$$

We wish to derive general expressions for X' and C_z' in terms of the wavenumber weight function $f(k,\kappa)$ in

$$\psi(r) = e^{im\phi} \int_0^k d\kappa \, f(k,\kappa) \, J_m(\kappa\rho) \, e^{iqz}, \qquad \kappa^2 + q^2 = k^2. \tag{5.16}$$

We set the wavefunction ψ in the formulae (5.13) and (5.14) equal to the expression (5.16), and its complex conjugate to

$$\psi^*(r) = e^{-im\phi} \int_0^k d\kappa' \, f(k,\kappa') \, J_m(\kappa'\rho) \, e^{-iq'z}, \qquad \kappa'^2 + q'^2 = k^2. \tag{5.17}$$

We substitute (5.16) and (5.17) into the formulae for X' and C_z'. In the case of X' we integrate over ϕ first and are left with integrals over κ, κ', and ρ. The integrands contain the Bessel functions $J_m(\kappa\rho)$, $J_m(\kappa'\rho)$ and their derivatives. The derivatives are removed by use of the first of the following sets of recurrence relations (Watson 1944, Section 2.12 [10])

$$2J_m'(\zeta) = J_{m-1}(\zeta) - J_{m+1}(\zeta), \qquad \frac{2m}{\zeta} J_m(\zeta) = J_{m-1}(\zeta) + J_{m+1}(\zeta). \tag{5.18}$$

Next we use the second of the set of recurrence relations to obtain terms proportional to ρ times products of pairs of Bessel functions. The reduced integrand of X' then has the ρ dependent term

$$\rho[J_{m-1}(\kappa\rho)\, J_{m-1}(\kappa'\rho) - J_{m+1}(\kappa\rho)\, J_{m+1}(\kappa'\rho)]. \tag{5.19}$$

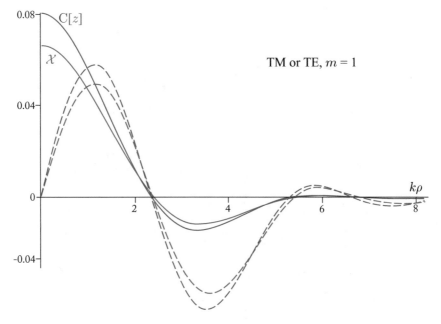

Figure 5.1: The chiral density $\chi(\rho, 0)$ and chiral current $C_z(\rho, 0)$ in the focal plane of the $m = 1$ TM or TE beams based on the wavefunction ψ_1. Also shown are $\rho\chi$ and ρC_z (dashed curves). The integrals of $\rho\chi$ and ρC_z over the range $0 \leq \rho < \infty$ are exactly zero.

Integration over ρ gives zero by Hankel's inversion formula (Watson 1944, Section 14.4 [10])

$$\int_0^\infty d\rho\, \rho\, J_m(\kappa\rho)\, J_m(\kappa'\rho) = \kappa^{-1}\delta(\kappa - \kappa') \qquad (\kappa, \kappa' > 0). \tag{5.20}$$

Hence *the chiral content per unit length of TM or TE beams is zero,* $X' = 0$, irrespective of the azimuthal winding number m.

The same method shows that *the total longitudinal component content of the chiral current is zero,* $C_z' = 0$. These results hold for all TM or TE beams based on wavefunctions of the form (5.16). The chiral density and the chiral current are not everywhere zero (except that $\chi = 0$, and C_ρ, $C_z = 0$ when $m = 0$, as is evident from ((5.13) and (5.14))), but their integrals over a slice of the beam at fixed z are zero, for any z. An analogous result holds for electromagnetic TM or TE *pulses*: for these the total chiral content $X = \int d^3r \chi = \int_0^\infty d\rho\, \rho \int_0^{2\pi} d\phi \int_{-\infty}^\infty dz\, \chi(\mathbf{r}, t)$ is zero (Lekner 2018, Section 2 [2]).

Figure 5.1 shows $\chi(\rho, 0)$ and $C_z(\rho, 0)$ in the focal plane of $m = 1$ TM or TE beams based on the wavefunction ψ_1 of Section 2.8. Let $\xi = k\rho$, $s = \sin \xi$, $c = \cos \xi$, $J_0 = J_0(\xi)$, $J_1 = J_1(\xi)$. The chiral density and longitudinal component of chiral current take the following forms

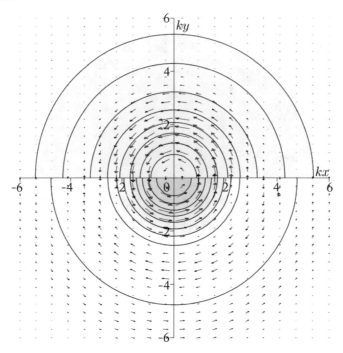

Figure 5.2: The chiral density and chiral current (lower half of figure, contours and arrows) compared with the energy density and azimuthal momentum density (upper half of figure). Drawn in the focal plane of the $m = 1$ TM or TE beams based on the wavefunction ψ_1. There is a reversal of sign in both $\chi(\rho, 0)$ and $C_\phi(\rho, 0)$.

in the focal plane:

$$k\chi = \xi^{-8} \left\{ 4\xi \left[\xi \left(\xi^2 - 21 \right) c - \left(8\xi^2 - 21 \right) s \right] J_0 - 4\xi \left(5\xi^2 - 42 \right) c J_1 - 4(\xi^4 - 19\xi^2 + 42)s \right\}$$
$$k^2 C_z = 4\xi^{-10} \left\{ -3\xi^6 J_0^2 - \xi^5 \left(\xi^2 - 12 \right) J_0 J_1 + 2\xi^4 \left(\xi^2 - 6 \right) J_1^2 \right.$$
$$\left. + \left(8\xi^4 - 63\xi^2 + 36 \right) c^2 + \xi(\xi^4 - 30\xi^2 + 72)sc - 5\xi^4 + 27\xi^2 - 36 \right\} . \qquad (5.21)$$

We see in Figure 5.1 that the chiral density and current have positive and negative parts, which exactly cancel in integrals over a beam cross-section, so that $X = \int d^3r \chi = 0$, $C_z' = \int d^3r C_z = 0$, as proved above. It may be verified that these integrals do indeed vanish when χ and C_z take the functional forms given in (5.21).

In Figure 5.2 we compare the focal plane chiral density and azimuthal current with the energy density and azimuthal momentum density, for the same beam. The azimuthal current density (we again set $\xi = k\rho$, $J_0 = J_0(\xi)$, $J_1 = J_1(\xi)$)

$$k^2 C_\phi(\rho, 0) = 4\xi^{-7} \left[\xi J_0 - 2J_1 \right] \left[\xi \left(\xi^2 - 8 \right) J_0 - 4 \left(\xi^2 - 4 \right) J_1 \right]. \qquad (5.22)$$

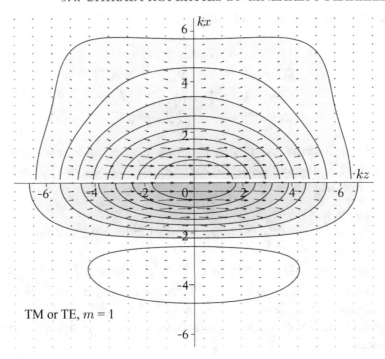

Figure 5.3: The chiral density and chiral current (lower half of figure, contours and arrows) compared with the energy density and momentum density (upper half of figure). Drawn in longitudinal section of the $m = 1$ TM or TE beams based on the wavefunction ψ_1. There is a reversal in sign of both $\chi(\rho, z)$ and $C_z(\rho, z)$ in the focal region.

The radial current density is zero.

Finally, in Figure 5.3 we compare the axial longitudinal sections of chiral density and current with the energy density and momentum density, which have already been shown in Figure 3.6. The reversals in sign of $\chi(\rho, z)$ and $C_z(\rho, z)$ were seen in the focal plane in Figure 5.1.

5.4   CHIRAL PROPERTIES OF 'LINEARLY POLARIZED' BEAMS

'Linearly polarized' beams were considered in Section 4.7. Those which are approximately linearly polarized along the x direction are based on the vector potential $A = [\psi, 0, 0]$. The prefactor $A_0 = k^{-1} E_0$ is again suppressed. The complex field amplitudes in rectangular coordinates are (we shall use subscripts to designate derivatives)

$$B = \left[0, \psi_z, -\psi_y\right], \qquad E = i\left[\psi_{xx} + k^2\psi, \ \psi_{xy}, \ \psi_{xz}\right]. \qquad (5.23)$$

The chiral density and current are found from (5.6):

$$\chi = k^{-1} Re\{\psi_z \psi_{xy}^* - \psi_y \psi_{xz}^*\} \tag{5.24}$$

$$C_x = -k^{-2} Im\{\psi_{xy} \psi_{xz}^* + k^2 \psi_y \psi_z^*\}$$
$$C_y = -k^{-2} Im\{\psi_{xz} [\psi_{xx}^* + k^2 \psi^*]\} \tag{5.25}$$
$$C_z = k^{-2} Im\{\psi_{xy} [\psi_{xx}^* + k^2 \psi^*]\}.$$

When converted to cylindrical coordinates, in which these expressions are more complicated, we can integrate over ϕ. The results are that the integrals of χ, C_ρ and C_z are all proportional to m (and thus vanish when $m = 0$)

$$k \int_0^{2\pi} d\phi \, \chi = m \frac{\pi}{\rho} Im \{\psi \psi_{\rho z}^* + \psi_\rho \psi_z^*\} = m \frac{\pi}{\rho} Im \{\partial_\rho (\psi \psi_z^*)\}. \tag{5.26}$$

Hence the integral $X' = \int d^2r \, \chi = \int_0^\infty d\rho \, \rho \int_0^{2\pi} d\phi \, \chi = -m\pi \, Im[\psi \psi_z^*]_{\rho=0}$. For $m > 0$ the wavefunction is zero on the beam axis $\rho = 0$. Hence the integral is zero, for any m. This result also follows by the methods of Section 5.3, as shown in Appendix 5A.

The only component of C not having m as factor when integrated over ϕ is the azimuthal one, for which we find

$$k^2 \int_0^{2\pi} d\phi \, C_\phi = \pi \, Im \{[\psi_{\rho\rho} + k^2\psi] \psi_{\rho z}^* + k^2 \psi_\rho \psi_z^*\}. \tag{5.27}$$

The longitudinal component of the chiral current, integrated over ϕ, is given by

$$k^2 \int_0^{2\pi} d\phi \, C_z = m \frac{\pi}{\rho^2} Re \left\{ \psi_{\rho\rho} (\rho\psi_\rho^* - \psi^*) - \psi_\rho \psi_\rho^* + (m^2 + 1)\rho^{-1}\psi \psi_\rho^* - m^2 \rho^{-2} \psi \psi^* \right\}$$
$$= m \frac{\pi}{\rho^2} Re \{(\psi_{\rho\rho} - \rho^{-1}\psi_\rho + m^2\rho^{-2}\psi)(\rho\psi_\rho^* - \psi^*)\}. \tag{5.28}$$

This is zero when $m = 0$, and in fact zero for all m, as shown in Appendix 5A. Hence

$$X' = \int d^2r \, \chi = 0, \qquad C_z' = \int d^2r \, C_z = 0 \qquad \text{('LP' beams, all } m). \tag{5.29}$$

The chiral density and the longitudinal chiral current are not (in general) zero everywhere, but their integrals over a slice of the beam at fixed z are zero, for any z.

For the 'LP' beam based on ψ_0, $\chi = 0$ in the focal plane, C_ρ averages to zero when integrated over the range of ϕ, and C_z is identically zero.

In the focal plane of the 'LP' beam based on ψ_1, C_ρ integrated over the range of ϕ is again zero. For the chiral density χ and the chiral current density C_z we find, with $\xi = k\rho$, $s =$

$\sin \xi$, $c = \cos \xi$, $J_0 = J_0(\xi)$, $J_1 = J_1(\xi)$ as before,

$$k \int_0^{2\pi} d\phi \, \chi = -4\pi\xi^{-8} \left\{ \left[\xi \left(\xi^2 - 21 \right) c - \left(8\xi^2 - 21 \right) s \right] \xi J_0 \right.$$
$$\left. - \left[5\xi \left(\xi^2 - 42 \right) c + \left(\xi^4 - 19\xi^2 + 42 \right) s \right] J_1 \right\} \qquad (5.30)$$

$$k^2 \int_0^{2\pi} d\phi \, C_z = 4\pi\xi^{-8} \left\{ 4\xi J_0 + \left(\xi^2 - 8 \right) J_1 \right\} \left\{ \xi \left(\xi^2 - 16 \right) J_0 - \left(6\xi^2 - 32 \right) J_1 \right\}.$$

It may be verified that the integrals $X' = \int d^2r \, \chi$, $C_z' = \int d^2r \, C_z$ vanish when $\int_0^{2\pi} d\phi \, \chi$ and $\int_0^{2\pi} d\phi \, C_z$ take the functional forms given in (5.30).

The identities $\int_0^\infty d\xi \xi$ {right-hand sides in (5.21), (5.30)} $= 0$ are just the $z = 0$ special cases of four families of null integrals, since analogues of (5.21) and (5.30) are valid at any z.

Figure 5.4 shows the energy, momentum and chiral densities for the 'LP' beam based on ψ_1, in its focal plane. The energy and momentum densities were seen previously in Figure 4.17. The logarithm of energy density was plotted there to put more emphasis outside the central focal region, and did not distinguish the focal plane local maxima, as does the present figure.

5.5   CHIRAL PROPERTIES OF SELF-DUAL MONOCHROMATIC BEAMS

Self-dual electromagnetic fields, considered in Section 3.6, have complex field amplitudes related by $\boldsymbol{E} = \pm i \boldsymbol{B}$, or $\boldsymbol{E}_r + i \boldsymbol{E}_i = \pm i(\boldsymbol{B}_r + i \boldsymbol{B}_i)$. Hence, for self-dual beams,

$$\boldsymbol{E}_r = \mp \boldsymbol{B}_i, \qquad\qquad \boldsymbol{E}_i = \pm \boldsymbol{B}_r. \qquad (5.31)$$

Substitution of (5.31) into (5.6) gives

$$\chi = \pm k \left(E_r^2 + E_i^2 \right) = \pm k \left(B_r^2 + B_i^2 \right), \qquad \boldsymbol{C} = 2k \left(\boldsymbol{E}_r \times \boldsymbol{E}_i \right) = 2k \left(\boldsymbol{B}_r \times \boldsymbol{B}_i \right). \qquad (5.32)$$

We compare these with the self-dual energy and momentum densities in (3.53) written in the form

$$u = \frac{1}{8\pi} \left(E_r^2 + E_i^2 \right) = \frac{1}{8\pi} \left(B_r^2 + B_i^2 \right), \qquad \boldsymbol{p} = \frac{1}{4\pi c} \left(\boldsymbol{E}_r \times \boldsymbol{E}_i \right) = \frac{1}{4\pi c} \left(\boldsymbol{B}_r \times \boldsymbol{B}_i \right). \qquad (5.33)$$

Hence *self-dual monochromatic fields have chiral density and chiral current strictly proportional to the energy density and c times the momentum density, with the same proportionality constant* ($\pm 8\pi k$ for the Gaussian units used here).

Another way of writing the densities is in terms of the complex field amplitudes. This gives

$$\chi(\boldsymbol{r}) = \pm k \, \boldsymbol{E} \cdot \boldsymbol{E}^* = \pm k \, \boldsymbol{B} \cdot \boldsymbol{B}^*, \qquad \boldsymbol{C}(\boldsymbol{r}) = ik \, \boldsymbol{E} \times \boldsymbol{E}^* = ik \, \boldsymbol{B} \times \boldsymbol{B}^*. \qquad (5.34)$$

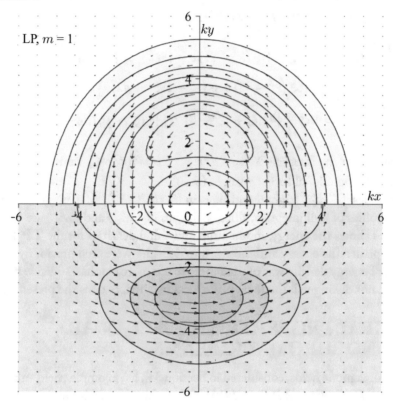

Figure 5.4: The chiral density and chiral current (lower half of figure, contours and arrows) compared with the energy density and transverse momentum density (upper half of figure). Drawn in the focal plane of the $m = 1$ 'LP' beam based on the wavefunction ψ_1. Note the reversal in sign of both $\chi(x, y, 0)$ and $C_\phi(x, y, 0)$. $\chi(0, 0, 0)$ is negative.

Comparison with the energy and momentum densities in (3.60) again gives (Lekner 2019 [4])

$$\chi = \pm (8\pi k) u, \qquad\qquad C = (8\pi k) c p. \qquad\qquad (5.35)$$

In electromagnetic pulses the $8\pi k$ factor also converts energy content to chiral content, but occurs inside the wavenumber integral (Lekner 2018a [2], Appendix B of Chapter 3 in Lekner 2018b [3]).

It was stated in Section 5.1 that the fact that E and B are eigenstates of curl for self-dual beams implies that such beams have maximal chirality. This is clear from the way in which fields and their curls enter into the general definitions (5.1) and (5.2). When $\nabla \times E = k E$ and $\nabla \times B = k B$ these become, for real fields, $\chi = k(E \cdot E + B \cdot B)$, $C = 2k E \times B$.

Self-duality of monochromatic fields implies *maximal chirality*, in the following sense: at any point in space the fact that the fields are eigenvectors of curl maximizes the local chiral density. The *global* maximum of both the energy and the chiral densities is in the focal region of beams. Different types of electromagnetic beams will have different forms of energy density.

The maximality of the magnitude the chiral density also follows from the work of Bliokh and Nori 2011 [1], who have shown by means of a Fourier representation of the electric and magnetic fields that the ratio of the chiral density to the energy density lies between $\pm 8\pi k$. Their equation (18) reads (using our definition of chiral density)

$$\frac{\chi}{8\pi \bar{u}} = \frac{\bar{u}_+ - \bar{u}_-}{\bar{u}_+ + \bar{u}_-}, \qquad \text{so} \qquad -1 \leq \frac{\chi}{8\pi \bar{u}} \leq 1. \tag{5.36}$$

The symbols $\bar{u}_+$, $\bar{u}_-$ denote energy densities for fields with positive and negative helicities, respectively, cycle-averaged. We have shown above that for self-dual fields the ratio has to be exactly $\pm 8\pi k$. This is a second proof of the theorem that *the chiral density is maximal for self-dual fields*.

Because of the equalities (5.35), there is no need to explore the chiral properties of self-dual beams that have already been discussed: their chiral density is given by the energy density, and their chiral current density by the momentum density. Hence the TM+iTE beams of Section 3.7, and the 'CP' beams of Sections 4.4–4.6 have known chiral density and chiral current, calculated from (5.35).

5A   APPENDIX: CHIRAL CONTENT OF 'LP' BEAMS

The chiral content and longitudinal chiral current content per unit length of beam are obtained by integrating over a section of the beam at constant z $\left(\int d^2 r = \iint_{-\infty}^{\infty} dx dy = \int_0^{\infty} d\rho\, \rho \int_0^{2\pi} d\phi \right)$:

$$X' = \int d^2 r\, \chi(r), \qquad C'_z = \int d^2 r\, C_z. \tag{5A.1}$$

We consider 'LP' beams in which the azimuthal dependence of ψ is $e^{im\phi}$. The chiral density and chiral current can be calculated from (5.23) and (5.24). There is ρ dependence and ϕ dependence in the integrands. We set

$$\psi^*(\rho, \phi, z) = e^{-im\phi} \int_0^k d\kappa\, f^*(k, \kappa)\, J_m(\kappa\rho)\, e^{-iqz}, \qquad \kappa^2 + q^2 = k^2$$

$$\psi(\rho, \phi, z) = e^{im\phi} \int_0^k d\kappa'\, f(k, \kappa')\, J_m(\kappa'\rho)\, e^{iq'z}, \qquad \kappa'^2 + q'^2 = k^2. \tag{5A.2}$$

We integrate over the azimuthal angle first, and then use the recurrence formulae (Watson 1944, Section 2.12 [10])

$$2J'_m(\zeta) = J_{m-1}(\zeta) - J_{m+1}(\zeta), \qquad \frac{2m}{\zeta} J_m(\zeta) = J_{m-1}(\zeta) + J_{m+1}(\zeta). \tag{5A.3}$$

After the elimination of derivatives, the chiral densities contain products of Bessel functions and inverse powers of ρ. The integrand of X' is the simplest, so we shall start with that. It contains the ρ dependent terms

$$\kappa J_{m+1}\left(\kappa\rho\right) J_m\left(\kappa'\rho\right) + \kappa' J_m\left(\kappa\rho\right) J_{m+1}\left(\kappa'\rho\right) - 2m\rho^{-1} J_m\left(\kappa\rho\right) J_m\left(\kappa'\rho\right). \qquad (5A.4)$$

We set $J_m\left(\zeta\right) = \frac{\zeta}{2m}\left[J_{m-1}\left(\zeta\right) + J_{m+1}\left(\zeta\right)\right]$ with $\zeta = \kappa\rho$ or $\kappa'\rho$. Thus X' is proportional to the integral

$$\int_0^\infty d\rho\,\rho\,\left\{J_{m+1}\left(\kappa\rho\right) J_{m+1}\left(\kappa'\rho\right) - J_{m-1}(\kappa\rho) J_{m-1}(\kappa'\rho)\right\}. \qquad (5A.5)$$

The integration over ρ can be performed using Hankel's inversion formula (Watson 1944, Section 14.4 [10]; Lekner 2001, Appendix A [5]), which may be written as

$$\int_0^\infty d\rho\,\rho\,J_m\left(\kappa\rho\right) J_m\left(\kappa'\rho\right) = \kappa^{-1}\delta\left(\kappa - \kappa'\right), \qquad \left(\kappa, \kappa' > 0\right). \qquad (5A.6)$$

The integration over ρ of products of Bessel functions of the same order thus selects $\kappa' = \kappa$, and thus also $q' = q$. The chiral content per unit length of the beam is thus zero, because of the cancellation in (5A.5).

Next, we consider C_z'. After elimination of derivatives, the chiral current longitudinal density contains Bessel functions and inverse powers of ρ, from ρ^{-1} to ρ^{-4}. Repeated use of $J_m\left(\zeta\right) = \frac{\zeta}{2m}\left[J_{m-1}\left(\zeta\right) + J_{m+1}\left(\zeta\right)\right]$ removes the inverse powers of ρ, and produces the following terms with equal order of Bessel function in the product (a constant common factor has been removed):

$$
\begin{aligned}
T_2 &= -\kappa^2\kappa'^2 J_{m+2}\left(\kappa\rho\right) J_{m+2}\left(\kappa'\rho\right) \\
T_1 &= m^{-1}(m+1)\kappa\kappa'(\kappa^2 + \kappa'^2) J_{m+1}\left(\kappa\rho\right) J_{m+1}\left(\kappa'\rho\right) \\
T_{-1} &= -m^{-1}(m-1)\kappa\kappa'(\kappa^2 + \kappa'^2) J_{m-1}\left(\kappa\rho\right) J_{m-1}\left(\kappa'\rho\right) \\
T_{-2} &= \kappa^2\kappa'^2 J_{m-2}\left(\kappa\rho\right) J_{m-2}\left(\kappa'\rho\right).
\end{aligned}
\qquad (5A.7)
$$

All these terms may be integrated by means of (5A.6). The sum of the integrals over ρ is

$$\int_0^\infty d\rho\,\rho\,T(\rho) = 4m^{-1}\kappa^3\delta\left(\kappa - \kappa'\right) \qquad (m \neq 0). \qquad (5A.8)$$

The terms with mixed order in the Bessel function products in C_z' are

$$
\begin{aligned}
R &= m^{-1}\kappa\kappa'[(m+1)\kappa'^2 - (m-1)\kappa^2] J_{m+1}\left(\kappa\rho\right) J_{m-1}\left(\kappa'\rho\right) \\
R' &= m^{-1}\kappa\kappa'[(m+1)\kappa^2 - (m-1)\kappa'^2] J_{m-1}\left(\kappa\rho\right) J_{m+1}\left(\kappa'\rho\right) \\
S &= \left(\kappa\kappa'\right)^2\{[J_{m-2}\left(\kappa\rho\right) - J_{m+2}\left(\kappa\rho\right)] J_m\left(\kappa'\rho\right) + J_m\left(\kappa\rho\right)\left[J_{m-2}\left(\kappa'\rho\right) - J_{m+2}\left(\kappa'\rho\right)\right]\}.
\end{aligned}
\qquad (5A.9)
$$

We note that the terms T_1, T_{-1}, R, R' are individually singular at $m = 0$, at which value their sum has a removable singularity, and is in fact zero. Also S is zero when $m = 0$. We saw in Section 5.4 that C_z' is proportional to m, which is confirmed by the fact that the terms containing the factor m^{-1} add to zero when $m \to 0$.

For the integration over ρ of $R + R' + S$ we need the singular integrals

$$U_m(\kappa, \kappa') = \int_0^\infty d\rho \, \rho \, J_{m-1}(\kappa\rho) \, J_{m+1}(\kappa'\rho) \tag{5A.10}$$

$$V_m(\kappa, \kappa') = \int_0^\infty d\rho \, \rho \, J_m(\kappa\rho) \, J_{m+2}(\kappa'\rho) = U_{m+1}(\kappa, \kappa'). \tag{5A.11}$$

In the U_m integral we set $J_{m-1}(\kappa\rho) = \frac{2m}{\kappa\rho} J_m(\kappa\rho) - J_{m+1}(\kappa\rho)$. This gives, on using (5A.6),

$$U_m(\kappa, \kappa') = -\kappa^{-1}\delta(\kappa - \kappa') + \frac{2m}{\kappa} \int_0^\infty d\rho \, J_m(\kappa\rho) \, J_{m+1}(\kappa'\rho). \tag{5A.12}$$

The integral remaining in (5A.12) is a special case of the Weber-Schafheitlin discontinuous integral (Watson 1944 Section 13.42 (8) [10]),

$$W_m(\kappa, \kappa') = \int_0^\infty d\rho \, J_m(\kappa\rho) \, J_{m+1}(\kappa'\rho) = \begin{cases} \dfrac{\kappa^m}{\kappa'^{m+1}}, & \kappa < \kappa' \\ \dfrac{1}{2\kappa}, & \kappa = \kappa' \\ 0, & \kappa > \kappa'. \end{cases} \tag{5A.13}$$

Note that $W_{m+1}(\kappa, \kappa') = (\kappa/\kappa') W_m(\kappa, \kappa')$, $W_{m-1}(\kappa, \kappa') = (\kappa'/\kappa) W_m(\kappa, \kappa')$. We also need integrals in which κ, κ' are interchanged, implicit in (5A.13).

Substitution of the delta function components of $U_m(\kappa, \kappa')$ into $\int_0^\infty d\rho\rho \, (R + R')$ cancels (5A.8). The remaining parts of $\int_0^\infty d\rho \, \rho \, (R + R' + S)$ may be expressed in terms of $W_m(\kappa, \kappa')$ and $W_m(\kappa', \kappa)$.

We shall list the terms separately (note that the delta function contributions have been omitted):

$$\begin{aligned} R &\to 2\kappa[(m+1)\kappa'^2 - (m-1)\kappa^2]W_m(\kappa', \kappa) \\ R' &\to 2\kappa'\left[(m+1)\kappa^2 - (m-1)\kappa'^2\right] W_m(\kappa, \kappa') \\ S &\to (\kappa\kappa')^2 \left\{(2m-2)\left[\kappa^{-1}W_{m-1}(\kappa, \kappa') + \kappa'^{-1}W_{m-1}(\kappa', \kappa)\right] \right. \\ &\left. \quad - (2m+2)\left[\kappa^{-1}W_{m+1}(\kappa, \kappa') + \kappa'^{-1}W_{m+1}(\kappa', \kappa)\right]\right\}. \end{aligned} \tag{5A.14}$$

From (5A.13) we see that

$$W_{m+1}(\kappa, \kappa') = \frac{\kappa}{\kappa'} W_m(\kappa, \kappa'), \qquad W_{m-1}(\kappa, \kappa') = \frac{\kappa'}{\kappa} W_m(\kappa, \kappa'). \tag{5A.15}$$

Hence

$$\begin{aligned}
S \to (2m - 2) & \left[\kappa'^3 W_m \left(\kappa, \kappa' \right) + \kappa^3 W_m \left(\kappa', \kappa \right) \right] \\
& - (2m + 2) \left[\kappa^2 \kappa' W_m \left(\kappa, \kappa' \right) + \kappa \kappa'^2 W_m \left(\kappa', \kappa \right) \right] \\
= -2\kappa' & \left[(m + 1) \kappa^2 - (m - 1) \kappa'^2 \right] W_m \left(\kappa, \kappa' \right) \\
& - 2\kappa \left[(m + 1) \kappa'^2 - (m - 1) \kappa^2 \right] W_m \left(\kappa', \kappa \right).
\end{aligned} \tag{5A.16}$$

Thus the terms arising from S annul those arising from R and R', and we have a grand cancellation to zero. The net longitudinal chiral current content per unit length of an 'LP' beam is zero: $C_z' = 0$, for any 'longitudinally polarized' beam.

5.7   CITED REFERENCES

[1] Bliokh, K. Y. and Nori, F. 2011. Characterizing optical chirality, *Physical Review A*, 83(021803). DOI: 10.1103/physreva.83.021803. 96, 105

[2] Lekner, J. 2018a. Chiral content of electromagnetic pulses, *Journal of Optics*, 20(105605):13. DOI: 10.1088/2040-8986/aadef5. 99, 104

[3] Lekner, J. 2018b. *Theory of Electromagnetic Pulses*, IOP Concise Physics (Bristol). DOI: 10.1088/978-1-6432-7022-7. 104

[4] Lekner, J. 2019. Chirality of self-dual electromagnetic beams, *Journal of Optics*, 21(035402):7. DOI: 10.1088/2040-8986/ab026f. 95, 104

[5] Lekner, J. 2001. TM, TE and "TEM" beam modes: Exact solutions and their problems, *Journal of Optics A: Pure Applied Optics*, 3:407–412. DOI: 10.1088/1464-4258/3/5/314. 106

[6] Lipkin, D. M. 1964. Existence of a new conservation law in electromagnetic theory, *Journal of Mathematical Physics*, 5:696–700. DOI: 10.1063/1.1704165. 95

[7] Lowry, T. M. 1964. *Optical Rotatory Power*, Dover, New York. DOI: 10.1038/125762a0. 95

[8] Tang, Y. and Cohen, A. E. 2010. Optical chirality and its interaction with matter, *Physical Review Letters*, 104(163901). DOI: 10.1103/physrevlett.104.163901. 95

[9] Tang, Y. and Cohen, A. E. 2011. Enhanced enantioselectivity in excitation of chiral molecules by superchiral light, *Science*, 332:333–336. DOI: 10.1126/science.1202817. 95

[10] Watson, 1944. *Theory of Bessel Functions*, Cambridge University Press. 98, 99, 105, 106, 107

5.8   ADDITIONAL REFERENCES (CHRONOLOGICAL ORDER)

[11] Kibble, T. W. B. 1965. Conservation laws for free fields, *Journal of Mathematical Physics*, 6:1022–1026. DOI: 10.1063/1.1704363.

[12] Fairlie, D. B. 1965. Conservation laws and invariance principles, *Il Nuovo Cimento*, 37:897–904. DOI: 10.1007/bf02773179.

[13] Candlin, D. J. 1965. Analysis of a new conservation law in electromagnetic theory, *Il Nuovo Cimento*, 37:1390–1395. DOI: 10.1007/bf02783348.

[14] Calkin, M. G. 1965. An invariance property of the free electromagnetic field, *American Journal of Physics*, 33:958–960. DOI: 10.1119/1.1971089.

[15] Barnes, A. 1977. Geometrical meaning of the curl operation when $A.curl\,A \neq 0$, *American Journal of Physics*, 45:371–372. DOI: 10.1119/1.10846.

[16] Thomsen, J. S. 1978. Comment on geometrical meaning of the curl operation when $A.curl\,A \neq 0$, *American Journal of Physics*, 46:684–685. DOI: 10.1119/1.11234.

[17] Sudbery, A. 1986. A vector Lagrangian for the electromagnetic field, *Journal of Physics A: Mathematical and General*, 19:L33–L36. DOI: 10.1088/0305-4470/19/2/002.

[18] McKelvey, J. P. 1990. The case of the curious curl, *American Journal of Physics*, 58:306–310. DOI: 10.1119/1.16161.

[19] McLaughlin, D. and Pironneau, O. 1991. Some notes on periodic Beltrami fields in Cartesian geometry, *Journal of Mathematical Physics*, 32:797–804. DOI: 10.1063/1.529373.

[20] Afanasiev, G. N. and Stepanovsky, Y. P. 1996. The helicity of the free electromagnetic field, *Il Nuovo Cimento*, 109A:271–279. DOI: 10.1007/BF02731014.

[21] Coles, M. M. and Andrews, D. L. 2012. Chirality and angular momentum in optical radiation, *Physical Review A*, 85(063810). DOI: 10.1103/physreva.85.063810.

[22] Andrews, D. L. and Coles, M. M. 2012. Measures of chirality and angular momentum in the electromagnetic field, *Optics Letters*, 37:3009–3011. DOI: 10.1364/ol.37.003009.

[23] Barnett, S. M., Cameron, R. P., and Yao, A. M. 2012. Duplex symmetry and its relation to the conservation of optical helicity, *Physics Review A*, 86:013845. DOI: 10.1103/physreva.86.013845.

[24] Cameron, R. P. and Barnett, S. M. 2012. Electro-magnetic symmetry and Noether's theorem, *New Journal of Physics*, 14(123019). DOI: 10.1088/1367-2630/14/12/123019.

[25] Fernandez-Corbaton, I., Zambrana-Puyalto, X., Tischler, N., and Molina-Terriza, G. 2012. Helicity and angular momentum: A symmetry-based framework for the study of light-matter interactions, *Physics Review A*, 86(042103). DOI: 10.1103/physreva.86.042103.

[26] Cameron, R. P., Barnett, S. M., and Yao, A. M. 2012. Optical helicity, optical spin and related quantities in electromagnetic theory, *New Journal of Physics*, 14(053050). DOI: 10.1088/1367-2630/14/5/053050.

[27] Bliokh, K., Bekshaev, A., and Nori, F. 2013. Dual electromagnetism: Helicity, spin, momentum and angular momentum, *New Journal of Physics*, 15(033026), Corrigendum 2018, *New Journal of Physics*, 18(089503). DOI: 10.1088/1367-2630/15/3/033026.

[28] Philbin, T. G. 2013. Lipkin's conservation law, Noether's theorem, and the relation to optical helicity, *Physics Review A*, 87(043843). DOI: 10.1103/physreva.87.043843.

[29] Fernandez-Corbaton, I., Zambrana-Puyalto, X., Tischler, N., Vidal, X., Juan, M. L., and Molina-Terriza, G. 2013. Electromagnetic duality symmetry and helicity conservation for macroscopic Maxwell's equations, *Physics Review Letters*, 111(060401). DOI: 10.1103/physrevlett.111.060401.

[30] Cameron, R. P., Barnett, S. M., and Yao, A. M. 2014. Optical helicity of interfering waves, *Journal of Modern Optics*, 61:25–31. DOI: 10.1080/09500340.2013.829874.

[31] Cameron, R. P., Götte, J. B., Barnett, S. M., and Yao, A. M. 2016. Chirality and the angular momentum of light, *Philosophical Transactions on Royal Society A*, 375(20150433). DOI: 10.1098/rsta.2015.0433.

[32] Crimin, F., Mackinnon, N., Götte, J. B., and Barnett, S. M. 2019. Optical helicity and chirality: Conservation and sources, *Applied Sciences*, 9(828):17. DOI: 10.3390/app9050828.

[33] Crimin, F., Mackinnon, N., Götte, J. B., and Barnett, S. M. 2019. On the conservation of helicity in a chiral medium, *Journal of Optics*, 21(094003):6. DOI: 10.1088/2040-8986/ab387c.

[34] Mackinnon, N. 2019. On the differences between helicity and chirality, *Journal of Optics*, 21(125405):7. DOI: 10.1088/2040-8986/ab4db9.

CHAPTER 6

Comparison of Electromagnetic Beams

6.1 ENERGY, MOMENTUM AND ANGULAR MOMENTUM PER UNIT LENGTH OF BEAMS

We have previously derived expressions for the energy, momentum and angular momentum per unit length of TM and TE, TM+iTE, 'CP' and 'LP' beams, given as integrals over the wavenumber weight function $f(k, \kappa)$ which determines a general beam wavefunction:

$$\psi(\rho, \phi, z) = e^{im\phi} \int_0^k d\kappa \, f(k,\kappa) \, J_m(\kappa\rho) \, e^{iqz}. \tag{6.1}$$

The following expressions, respectively (3.33), (3.61), (4B.15), and (4.47), summarize the results:

$$\begin{bmatrix} U' \\ cP'_z \\ cJ'_z \end{bmatrix} = \frac{E_0^2}{4k^3} \int_0^k d\kappa |f(k,\kappa)|^2 \kappa \begin{bmatrix} k \\ q \\ m \end{bmatrix} \qquad \text{(TM or TE)} \tag{6.2}$$

$$\begin{bmatrix} U' \\ cP'_z \\ cJ'_z \end{bmatrix} = \frac{E_0^2}{2k^3} \int_0^k d\kappa |f(k,\kappa)|^2 \kappa \begin{bmatrix} k \\ q \\ m + \kappa^2/2kq \end{bmatrix} \qquad \text{(self-dual TM+iTE)} \tag{6.3}$$

$$\begin{bmatrix} U' \\ cP'_z \\ cJ'_z \end{bmatrix} = \frac{E_0^2}{8k^3} \int_0^k d\kappa |f(k,\kappa)|^2 (k+q)^2 \kappa^{-1} \begin{bmatrix} k \\ q \\ m + 1 + \frac{\kappa^2}{2kq} \end{bmatrix} \qquad \text{(self-dual 'CP')} \tag{6.4}$$

$$\begin{bmatrix} U' \\ cP'_z \\ cJ'_z \end{bmatrix} = \frac{E_0^2}{8k^3} \int_0^k d\kappa |f(k,\kappa)|^2 (k^2 + q^2) \kappa^{-1} \begin{bmatrix} k \\ q \\ m \end{bmatrix} \qquad \text{('LP').} \tag{6.5}$$

The equations for the TM or TE, and for the 'LP' beams, may be interpreted in terms of the light quantum: these beams can be viewed as a superposition of photons, each with energy $\hbar ck$, z component of momentum $\hbar q$, and z component of angular momentum $\hbar m$.

All the beam types share this identification of energy with $\hbar ck$, z component of momentum with $\hbar q$. However, the angular momentum content expression for the self-dual TM+iTE and 'CP' beams differs from the TM/TE and 'LP' beam case, where it was simply $\hbar m$ per photon. For the TM+iTE beams this becomes $\hbar \left(m + \frac{\kappa^2}{2kq}\right)$, a function of the wavenumber components $\kappa = k_x$, $q = k_z$. For the 'CP' beams the angular momentum per photon becomes $\hbar \left(m + 1 + \frac{\kappa^2}{2kq}\right)$. These expressions differ from the TM, TE, and 'LP' beams, where it was simply $\hbar m$. A possible interpretation is to associate 'spin' with the extra terms. Spin-orbital decomposition is discussed in Section 2 of Allen, Barnett and Padgett 2003 [1], also in Torres and Torner 2011 [13], and Andrews and Babiker 2013 [4].

Even though the angular momentum density of a classical electromagnetic field is all orbital since the angular momentum density is $j = r \times p$, one may ask whether it is useful to separate out the component of j_z proportional to m as 'orbital' and treat the remainder as 'spin.' Such a split is certainly useful in the case of planetary motion, where for example we distinguish between the angular momentum of the Earth in its orbit around the Sun, and the angular momentum of its rotation about the north-south axis.

Quantum theory uniquely associates orbital angular momentum $\hbar m$ (about the z axis) with a *scalar* field wavefunction that has its azimuthal dependence in the factor $e^{im\phi}$, since in quantum mechanics $L_z = x p_y - y p_x = -i\hbar \left(x\partial_y - y\partial_x\right) = -i\hbar\partial_\phi$, so that $L_z e^{im\phi} = \hbar m e^{im\phi}$, and the wavefunction is an eigenstate of L_z with eigenvalue $\hbar m$. Quantum particle beams based on a wavefunction with azimuthal dependence $e^{im\phi}$ have z component of angular momentum proportional to m (Lekner 2004a [6]). Likewise, sound propagation in an isotropic fluid is represented by a scalar field, and acoustic beams based on a wavefunction with azimuthal dependence $e^{im\phi}$ have z component of angular momentum proportional to m, with $\omega J_z' = m U'$ (Lekner 2006 [8]).

One problem arising with identifying orbital angular momentum components of *vector* fields is that phase quadrature between (for example) the x and y components of the field gives a phase factor $e^{i\phi}$, without any azimuthal dependence in the wavefunction. We saw an example in Section 4.4, where the vector potential (4.27) and the derived fields (4.28) have the phase factor $e^{i\phi}$, in addition to any azimuthal dependence in the wavefunction. This is the origin of the $m + 1$ term in the integral giving J_z' in the self-dual 'CP' beams.

Rather than continue the discussion of the orbital-spin separation, let us restate the *proven* relations between the angular momentum and energy contents per unit length of beams:

$$\omega J_z' = m U' \qquad \text{(TM, TE, and 'LP' beams)}$$
$$\omega J_z' = m U' + \text{positive term} \qquad \text{(self-dual TM+iTE beams)} \tag{6.6}$$
$$\omega J_z' = (m + 1) U' + \text{positive term} \qquad \text{(self-dual 'CP' beams)}.$$

These relations hold for all possible causal beams, namely those based on solutions of the wave equation of the form (6.1). As noted above, the relation $\omega J_z' = m U'$ holds also for acoustic beams, and is consistent with these beams being composed of vibrational quanta (phonons) of energy $\hbar c k = \hbar \omega$ and angular momentum $\hbar m$.

In the next Section we shall compare beam properties, as functions of the azimuthal winding number m. The simplest assumption (Lekner 2004b [7]) is to keep the same wavenumber weight function: $f_m = f_0$. When that holds, all the $m > 0$ derived beams have (remarkably) the same energy and momentum per unit length of beam as the $m = 0$ beam, despite having very different wavefunctions, with $J_0(\kappa\rho)$ replaced by $e^{im\phi} J_m(\kappa\rho)$ in the wavefunction integrands. The angular momentum and energy are related by $\omega J_z' = mU'$ for TE, TM, and 'LP' beams, as always. The only dependence on the weight function $f(k, \kappa)$ is in the angular momentum and energy relations for TM+iTE, and 'CP' beams, because of the extra term $\kappa^2/2kq$ in the integrands of (6.3) and (6.4). This extra term is independent of m if f is independent of m, and for the proto-beam $f = f_0 = 2\kappa/k^2$ we find, from (6.3) and (6.4), respectively,

$$\omega J_z' = \left(m + \frac{16}{15}\right) U' \qquad \text{(TM+iTE beams)}$$

$$\omega J_z' = \left(m + 1 + \frac{39}{85}\right) U' \qquad \text{('CP' beams)}.$$

In the following Sections we shall not assume $f_m = f_0$, but rather the relation (6.8), which changes the wavenumber weight function by the factor κ/k for each integer increase in m.

6.2 TIGHTLY-FOCUSED BEAMS COMPARED

In the defining Equation (6.1) the weight function $f(k, \kappa)$ in general depends on the azimuthal winding number m. We saw in Section 2.8 that repeated application of the operator $\partial_x + i\partial_y = e^{i\phi}\left(\partial_\rho + i\rho^{-1}\partial_\phi\right)$ raises the value of m by one, since

$$e^{i\phi}\left(\partial_\rho + i\rho^{-1}\partial_\phi\right) e^{im\phi} J_m(\kappa\rho) = -\kappa e^{i(m+1)\phi} J_{m+1}(\kappa\rho). \tag{6.7}$$

A possible choice for all the beams with $m > 0$ derived from a $m = 0$ weight function $f_0(k, \kappa)$ is thus

$$f_m(k, \kappa) = (\kappa/k)^m f_0(k, \kappa). \tag{6.8}$$

This relation holds for all beam families of the form (6.1). We have inserted the factor k^{-m} to keep the dimensionality of f_m the same as that of f_0 (a length).

When $m = 0$ the Porras beam and the proto-beam of Section 2.5, respectively, take the forms (we normalize ψ to unity at the origin)

$$\psi_{Porras}(\rho, z) = \frac{a}{k J_1(ka)} \int_0^k d\kappa \, \kappa \, J_0(\kappa a) J_0(\kappa\rho) e^{iqz}, \qquad J_0(ka) = 0 \tag{6.9}$$

$$\psi_0(\rho, z) = \frac{2}{k^2} \int_0^k d\kappa \, \kappa \, e^{iqz} J_0(\kappa\rho) = \frac{2}{k^2} \int_0^k dq \, q \, e^{iqz} J_0(\kappa\rho). \tag{6.10}$$

The proto-beam weight function is $f_0(k, \kappa) = 2\kappa k^{-2}$. For the Porras family

$$f_{P0} = \frac{\kappa a \, J_0(\kappa a)}{k J_1(ka)}, \qquad J_0(ka) = 0. \tag{6.11}$$

All the integrals arising in U', P'_z, J'_z for the beam families derived from the proto-beam may be evaluated in closed form. Hence we shall use the proto-beam weight functions in the comparisons. In the following we omit the factor $A_0^2 = E_0^2/k^2$ (of dimension energy/length) for simplicity.

For the TM or TE beams, and also for the 'LP' beams, the equality $ck J'_z = \omega J'_z = m U'$ holds, so we shall not list the J'_z values. We have previously evaluated the $m = 0, 1$ TM values, in (3.34) and (3.40). The general results for the TM or TE beams are:

$$U' = \frac{1}{2(m+2)}, \qquad c P'_z = \frac{2^{m+1}(m+1)!}{(2m+5)!!} \qquad \text{(TM or TE).} \qquad (6.12)$$

The ratio $c P'_z/U'$ is $\frac{8}{15} \approx 0.533$ for $m = 0$, and decreases steadily with m. The leading asymptotic term of the ratio at large m is $\frac{1}{2}\sqrt{\pi/m}$.

For the linearly polarized family of beams we find

$$U' = \frac{m+3}{4(m+2)(m+1)}, \qquad c P'_z = \frac{2^m(m+4)m!}{(2m+5)!!} \qquad \text{('LP').} \qquad (6.13)$$

The ratio $c P'_z/U'$ is $32/45 \approx 0.711$ for $m = 0$, and again decreases monotonically with m. The asymptotic leading term is the same as for the TM and TE beams. The angular momentum per unit length is strictly proportional to the energy per unit length: $ck J'_z = m U'$.

Next we look at the self-dual TM+iTE family based on the proto-beam. For the TM+iTE family we know from Appendix 3B that the energies and momenta per unit length of the beam are just twice those of the TM or TE beams. For the angular momentum content per unit length we find a non-zero value when $m = 0$, $ck J'_z = 8/15$. As m increases the ratio of $ck J'_z$ to U' increases, becoming proportional to m for large m. For $m = 0, 1$ the ratio $ck J'_z/U'$ takes the values 16/15 and 83/35, in agreement with (3.64) and (3.65), respectively. The general expression is

$$ck J'_z/U' = m + 32 \frac{4^m(m+2)(m+2)!(m+3)!}{(2m+6)!} \qquad \text{(TM+iTE).} \qquad (6.14)$$

Finally, the self-dual 'CP' family based on the proto-beam. We find

$$U' = \frac{m+3}{4(m+2)(m+1)} + 2\frac{4^m(m+1)!m!}{(2m+3)!}$$

$$c P'_z = \frac{1}{2(m+2)(m+1)} + 4\frac{4^m(m+4)(m+2)!m!}{(2m+5)!} \qquad \text{('CP')} \qquad (6.15)$$

$$ck J'_z = \frac{m+4}{4(m+2)} + 4\frac{4^m(3m+8)(m+2)!(m+1)!}{(2m+5)!}.$$

The $m = 0$ and $m = 1$ values given in (4.32) and (4.38), respectively, agree with these expressions. The ratio $c P'_z/U'$ is $62/85 \approx 0.729$ for $m = 0$, and decreases steadily with m. The leading asymptotic term of the ratio at large m is again $\frac{1}{2}\sqrt{\pi/m}$. For the ratio $ck J'_z/U'$ the $m = 0$ and

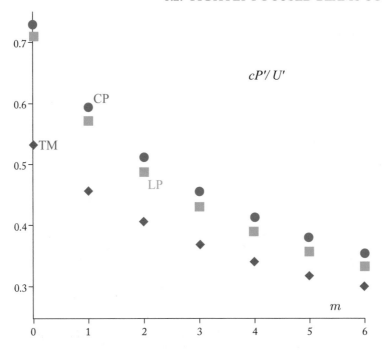

Figure 6.1: The variation of the cP_z'/U' ratios with the azimuthal winding number m, for the tightly focused beams derived from the proto-beam ψ_0 according to (6.8). The TM values shown apply also to TE and TM+iTE beams (see text).

$m = 1$ 'CP' values are 124/85, 39/14. For large m both the TM+iTE and the 'CP' beam families have the asymptotic form

$$ck J_z'/U' = m + \frac{1}{2}\sqrt{\pi m} + O(1) \qquad \text{(TM+iTE, 'CP').} \qquad (6.16)$$

Figure 6.1 shows how the cP_z'/U' ratios vary with the azimuthal winding number m. TE beams have the same energy and momentum densities as TM beams, and TM+iTE beams have different energy and momentum densities but as shown in Appendix 4B they have twice the energy and momentum per unit length of the beam compared with the TM and TE beams (for the same beam wavefunction). Hence the TM points for the cP_z'/U' ratios on the graph apply also to TE and TM+iTE beams.

We see form Figure 6.1 that for all the beam types the cP_z'/U' ratios decrease with m. The textbook beam has electric and magnetic fields orthogonal and transverse to the propagation direction. The energy and momentum densities have the ratio c for their moduli. This is only possible for plane waves (transversely infinite beams): see Theorems 4.1 and 4.4 of Appendix 4A. The more a beam departs from the plane wave idealization, the more the energy to momentum

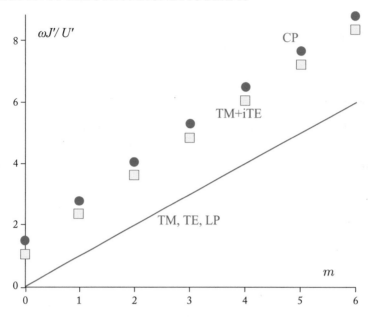

Figure 6.2: $ck J'_z/U' = \omega J'_z/U'$ ratios as functions of the azimuthal winding number m, for beams based on the proto-beam wavefunction. For the TM, TE, and LP beams $\omega J'_z/U' = m$, but for the self-dual TM+iTE, and 'CP' beams it is larger. For both the TM+iTE beams and the 'CP' beams the ratio has the large m asymptotic form $ck J'_z/U' = m + \frac{1}{2}\sqrt{\pi m} + O(1)$.

ratio differs from c. As the azimuthal winding number increases, beams of all types differ more and more from the plane wave, and thus the $c P'_z/U'$ ratios decrease with m. The maximum ratios at $m = 0$ are all less than unity, in accord with the energy-momentum inequality of Section 1.4.

Figure 6.2 shows the dependence on m of the $ck J'_z/U'$ ratios for the various beam families. For the TM, TE and 'LP' beams this ratio is simply m, represented by the straight line of unit slope. For the self-dual TM+iTE and 'CP' beams the ratio is larger than m, asymptotically $m + \frac{1}{2}\sqrt{\pi m} + O(1)$.

6.3   VARIABLE TIGHTNESS OF FOCUS

We wish to examine how the various electromagnetic beams, specifically the TM or TE, TM+iTE, 'LP,' and 'CP' beam families, vary as the tightness of focus is relaxed from that of the proto-beam used in the previous Section. We have seen in Section 2.5 that beam wavefunctions such as ψ_0 may be generalized by means of a complex shift along the propagation direction, $z \rightarrow z - ib$. Applied to the proto-beam, this gives the wavefunction (2.27) whose properties were explored in Section 2.6:

$$\psi_b\left(\rho, z\right) = \frac{b^2}{e^{kb}\left(kb-1\right)+1} \int_0^k d\kappa\,\kappa\,e^{qb+iqz} J_0\left(\kappa\rho\right)$$

$$= \frac{b^2}{e^{kb}\left(kb-1\right)+1} \int_0^k dq\,q\,e^{qb+iqz} J_0\left(\kappa\rho\right). \tag{6.17}$$

$(\kappa^2 + q^2 = k^2)$. The prefactor in (6.17) normalizes the wavefunction to unity at the origin $\rho = 0$, $z = 0$. The Carter (1973) [5] wavefunction ψ_C was given in (2.26), which we also normalize to unity at the origin:

$$\psi_C\left(\rho, z\right) = \frac{b/k}{1-e^{-kb/2}} \int_0^k d\kappa\,\kappa\,e^{-b\kappa^2/2k+iqz} J_0\left(\kappa\rho\right)$$

$$= \frac{b/k}{e^{kb/2}-1} \int_0^k dq\,q\,e^{bq^2/2k+iqz} J_0\left(\kappa\rho\right). \tag{6.18}$$

The beam wavefunctions ψ_b and ψ_C are compared in Figures 1–4 of Andrejic and Lekner 2017 [3]. Topological differences arise as the dimensionless parameter kb approaches 5, when some of the Carter wavefunction zeros move off the focal plane. As kb tends to zero the wavefunctions ψ_b and ψ_C have the proto-beam as their confluent form:

$$\psi_0\left(\rho, z\right) = \frac{2}{k^2} \int_0^k d\kappa\,\kappa\,e^{iqz} J_0\left(\kappa\rho\right) = \frac{2}{k^2} \int_0^k dq\,q\,e^{iqz} J_0\left(\kappa\rho\right). \tag{6.19}$$

Both ψ_C and ψ_b give families of electromagnetic beams characterized by the length b as well as by the wavenumber k. The focal region of the families based on the proto-beam is both longitudinally and transversely of order k^{-1} in extent, since k^{-1} is the only length in the problem. For ψ_C and ψ_b and other beam wavefunctions depending on a length b, we now have two lengths as parameters. In the Gaussian beam of Section 2.2, the minimum beam width (in the focal plane) is of order $\sqrt{b/k}$, while the longitudinal extent is of order b. The dimensionless parameter kb also determines the divergence angle of the beams. For ψ_b the divergence angle was shown in Section 2.6 to be (for large kb) $\theta \approx \sqrt{2/kb}$, the same as for the Gaussian beam. Andrejic (2018) [2] has shown that as $kb \to 0$ the proto-beam divergence angle is $\pi/4$. This is also the case for electromagnetic pulses which are superpositions of the proto-beam form (Lekner 2018, 2018b Section 3.8 [10, 11]). For $kb = 6$ the scalar ψ_b beam divergence angle is approximately $\pi/6$.

Since the weight function for ψ_b is expressed in terms of $q = \sqrt{k^2 - \kappa^2}$, it is more convenient to carry out the integrations over q. From (6.17) the $m = 0$ weight function for ψ_b is

$$f_0\left(k, \kappa\right) = \frac{b^2 \kappa e^{qb}}{e^{kb}\left(kb-1\right)+1}, \qquad \kappa = \sqrt{k^2 - q^2}. \tag{6.20}$$

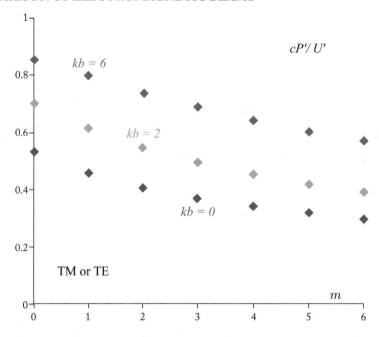

Figure 6.3: Ratio of momentum to energy per unit length of beam, shown for the TM or TE beams based on ψ_b, for variable m and $kb = 0, 2, 6$.

Higher azimuthal number beams are obtained, by (6.8), from the weight functions $f_m(k, \kappa) = (\kappa/k)^m f_0(k, \kappa)$.

We begin with the TM and TE beams. Using $qdq + \kappa d\kappa = 0$ we rewrite (6.2) as

$$
\begin{bmatrix} U' \\ cP'_z \\ cJ'_z \end{bmatrix} = \frac{E_0^2}{4k^3} \int_0^k dq |f(k, \kappa)|^2 q \begin{bmatrix} k \\ q \\ m \end{bmatrix} \qquad \text{(TM or TE)}. \qquad (6.21)
$$

Figure 6.3 shows how the momentum to energy ratio cP'_z/U' varies with the beam focal size parameter kb. As kb increases, the per unit length of beam momentum (times c) content to the energy content comes closer to the plane-wave value of unity.

We shall give the results in full for $m = 0$ beams. Those for higher m are similar but progressively more complicated. The angular momentum per unit length of beam is omitted, since it equals mU'/ω. As before we suppress the factor $E_0^2 k^{-2}$, and set $kb = \beta$.

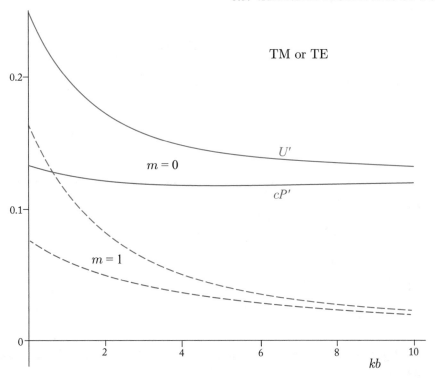

Figure 6.4: Energy and momentum per unit length for the TM or TE beams based on ψ_b with $m = 0,\ 1$. Since $\omega J'_z = mU'$ for TM or TE beams, the dashed blue curve represents both U' and $\omega J'_z$ for $m = 1$.

$$U' = \frac{\left(4\beta^2 - 6\beta + 3\right) e^{2\beta} + 2\beta^2 - 3}{32\left[(\beta - 1)\, e^\beta + 1\right]^2},$$

$$cP'_z = \frac{\left(2\beta^3 - 5\beta^2 + 6\beta - 3\right) e^{2\beta} - \beta^2 + 3}{16\beta\left[(\beta - 1)\, e^\beta + 1\right]^2} \qquad \text{(TM or TE, } m = 0). \qquad (6.22)$$

As $\beta = kb \to 0$ these results reproduce the $m = 0$ values in (6.12), namely $U' \to 1/4$, $cP'_z \to 2/15$. For large kb the asymptotic forms are

$$U' = \frac{1}{8} + \frac{1}{16\beta} + O\left(\beta^{-2}\right), \quad cP'_z = \frac{1}{8} - \frac{1}{16\beta} + O\left(\beta^{-2}\right) \qquad \text{(TM or TE, } m = 0).$$

$$(6.23)$$

Figure 6.4 shows the variation of U' and cP'_z with kb, for $m = 0$ and $m = 1$. In both cases the ratio cP'_z/U' tends to unity minus a term of order β^{-1}. This behavior is universal for

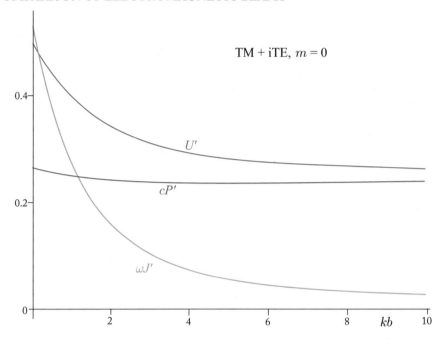

Figure 6.5: Energy, momentum and angular momentum per unit length for the $m = 0$ TM+iTE beams based on ψ_b.

all the beam types and for all azimuthal winding numbers. That $cP'_z/U' \leq 1$ is a consequence of the energy-momentum inequality of Section 1.4.

The TM+iTE beams have twice the energy and momentum per unit length of the TM beams, as shown in Appendix 3B. The angular momentum is however not just mU'/ω, as for the TM or TE beams. We rewrite (6.3) as

$$
\begin{bmatrix} U' \\ cP'_z \\ cJ'_z \end{bmatrix} = \frac{E_0^2}{2k^3} \int_0^k dq |f(k,\kappa)|^2 q \begin{bmatrix} k \\ q \\ m + \kappa^2/2kq \end{bmatrix} \qquad \text{(TM+iTE).} \qquad (6.24)
$$

For the $m = 0$ beam we find

$$
\omega J'_z = \frac{(4\beta^2 - 6\beta + 3) e^{2\beta} - 2\beta^4 + 2\beta^2 - 3}{16\beta[(\beta - 1) e^\beta + 1]^2} \qquad \text{(TM+iTE, } m = 0\text{).} \qquad (6.25)
$$

The leading terms at small and large β are

$$
\omega J'_z = \frac{8}{15} - \frac{17}{45}\beta + O\left(\beta^2\right), \qquad \omega J'_z = \frac{1}{4\beta} + \frac{1}{8\beta^2} + O\left(\beta^{-2}\right). \qquad (6.26)
$$

U', cP'_z, and ckJ'_z for the $m = 0$ TM+iTE beams are shown in Figure 6.5.

For the 'LP' beams, (6.5) rewritten as an integral over q reads

$$\begin{bmatrix} U' \\ cP'_z \\ cJ'_z \end{bmatrix} = \frac{E_0^2}{8k^3} \int_0^k dq |f(k,\kappa)|^2 (k^2 + q^2) q\kappa^{-2} \begin{bmatrix} k \\ q \\ m \end{bmatrix} \qquad \text{('LP').} \qquad (6.27)$$

The energy and momentum per unit length of the beam evaluate to

$$U' = \frac{(8\beta^3 - 8\beta^2 + 6\beta - 3) e^{2\beta} + 2\beta^2 + 3}{64[(\beta - 1) e^\beta + 1]^2} \qquad \text{('LP,' } m = 0)$$

$$cP'_z = \frac{(4\beta^4 - 6\beta^3 + 7\beta^2 - 6\beta + 3) e^{2\beta} - \beta^2 - 3}{32\beta [(\beta - 1) e^\beta + 1]^2}. \qquad (6.28)$$

The angular momentum per unit length of 'LP' beams is $J'_z = mU'/\omega$. As $\beta = kb \to 0$ these results reproduce the $m = 0$ values in (6.13), $U' \to 3/8$, $cP'_z \to 4/15$. For large kb the asymptotic forms are

$$U' = \frac{\beta}{8} + \frac{1}{8} + O(\beta^{-1}), \qquad cP'_z = \frac{\beta}{8} + \frac{1}{16} + O(\beta^{-1}) \qquad \text{('LP,' } m = 0). \qquad (6.29)$$

U' and cP'_z for the $m = 0$ and $m = 1$ 'LP' beams are shown in Figure 6.6. Since $ckJ'_z = mU'$ the angular momentum for the $m = 1$ beams is $J'_z = U'/\omega$.

Finally, we give the results for 'CP' beams. The relations (6.4) transcribe to

$$\begin{bmatrix} U' \\ cP'_z \\ cJ'_z \end{bmatrix} = \frac{E_0^2}{8k^3} \int_0^k dq |f(k,\kappa)|^2 (k + q)^2 q\kappa^{-2} \begin{bmatrix} k \\ q \\ m + 1 + \frac{\kappa^2}{2kq} \end{bmatrix} \qquad \text{('CP').} \qquad (6.30)$$

For the $m = 0$ 'CP' family we find

$$U' = \frac{(16\beta^3 - 16\beta^2 + 10\beta - 3) e^{2\beta} + 2\beta^2 - 4\beta + 3}{64[(\beta - 1) e^\beta + 1]^2}$$

$$cP'_z = \frac{(8\beta^4 - 12\beta^3 + 13\beta^2 - 9\beta + 3) e^{2\beta} - \beta^2 + 3\beta - 3}{32\beta [(\beta - 1) e^\beta + 1]^2} \qquad \text{('CP,' } m = 0) \qquad (6.31)$$

$$ckJ'_z = \frac{(16\beta^4 - 8\beta^3 - 2\beta^2 + 6\beta - 3) e^{2\beta} - 2\beta^4 + 4\beta^3 - 4\beta^2 + 3}{64\beta^2 [(\beta - 1) e^\beta + 1]^2}.$$

As $\beta = kb \to 0$ these results reproduce the $m = 0$ values in (6.15), namely $U' \to 17/24$, $cP'_z \to 31/60$ and $\omega J'_z \to 8/15$. For large kb the asymptotic forms are

$$U' = \frac{\beta}{4} + \frac{1}{4} + O(\beta^{-1}), \qquad cP'_z = \frac{\beta}{4} + \frac{1}{8} + O(\beta^{-1}).$$

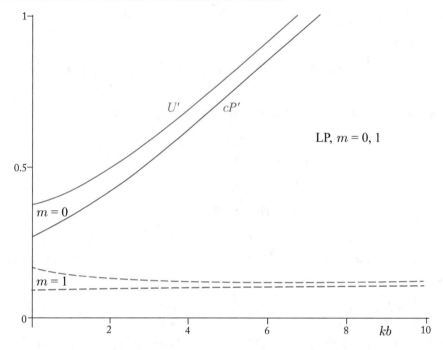

Figure 6.6: Energy and momentum per unit length for the $m = 0$ and $m = 1$ 'LP' beams based on ψ_b. We omit the angular momentum since $\omega J'_z = mU'$ for 'LP' beams.

$$ ck J'_z = \frac{\beta}{4} + \frac{3}{8} + O\left(\beta^{-1}\right) \qquad (\text{'CP'}, \, m = 0). \qquad (6.32) $$

U', $c P'_z$, and $ck J'_z$ for the $m = 0$ and $m = 1$ 'CP' beams are shown in Figure 6.7.

Summary

In this chapter, which is based on Lekner (2020) [12], we have compared the energy, momentum and angular momentum contents per unit length of various types of electromagnetic beams. The comparisons were made for beams based on the same scalar wavefunction. Two parameters were varied: the azimuthal winding number m, and the degree of tightness of focus, given by the dimensionless parameter kb.

When the wavenumber weight function has an extra factor of κ/k for each integer increase in m, the beams become less 'paraxial' as the winding number m gets larger. As m increases, the ratio of c times momentum to energy per unit length of beam decreases monotonically, and the ratio of ω times angular momentum to energy per unit length of beam increases. The increase is exactly proportional to m in the case of TM, TE, and 'LP' beams, and approximately so for the self-dual TM+iTE and 'CP' beams.

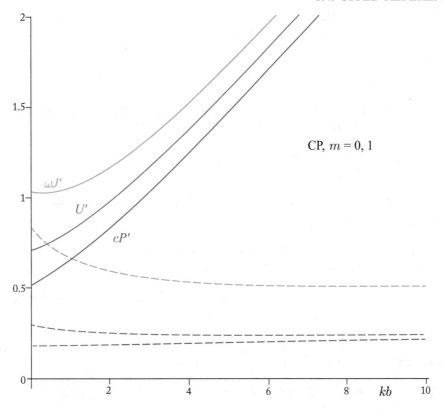

Figure 6.7: Energy, momentum and angular momentum per unit length for the $m = 0$ and $m = 1$ 'CP' beams based on ψ_b. The $m = 0$ curves are solid, the $m = 1$ curves dashed.

As the tightness of focus is loosened (by increasing the parameter kb), the beams become more plane-wave like, and the ratio of c times momentum to energy per unit length of beam tends to unity, for all beam types.

6.4   CITED REFERENCES

[1] Allen, L., Barnett, S. M., and Padgett, M. J. (Eds.) 2003. *Orbital Angular Momentum*, IOP Bristol. DOI: 10.1201/9781482269017. 112

[2] Andrejic, P. 2018. Convergent measure of focal extent, and largest peak intensity for non-paraxial beams, *Journal of Optics*, 20(075610):10. DOI: 10.1088%2F2040-8986%2Faaca6b. 117

[3] Andrejic, P. and Lekner, J. 2017. Topology of phase and polarization singularities in focal regions, *Journal of Optics*, 19(105609):8. DOI: 10.1088/2040-8986/aa895d. 117

[4] Andrews, D. L. and Babiker, M. 2013. *The Angular Momentum of Light*, Cambridge University Press. DOI: 10.1017/cbo9780511795213. 112

[5] Carter, W. H. 1973. Anomalies in the field of a Gaussian beam near focus, *Optics Communications*, 64:491–495. DOI: 10.1016/0030-4018(73)90012-6. 117

[6] Lekner, J. 2004a. Invariants of atom beams, *Journal of Physics B: Atomic and Molecular Physics*, 7:1725–1736. DOI: 10.1088/0953-4075/37/8/013. 112

[7] Lekner, J. 2004b. Invariants of three types of generalized Bessel beams, *Journal of Optics A: Pure Applied Optics*, 6:837–843. DOI: 10.1088/1464-4258/6/9/004. 113

[8] Lekner, J. 2006. Acoustic beams with angular momentum, *Journal of the Acoustical Society of America*, 120:3475–3478. DOI: 10.1121/1.2360420. 112

[9] Lekner, J. 2016. Tight focusing of light beams: A set of exact solutions, *Proc. of the Royal Society A*, 472(20160538):17. DOI: 10.1098/rspa.2016.0538.

[10] Lekner, J. 2018a. Electromagnetic pulses, localized and causal, *Proc. of the Royal Society A*, 474(20170655):17. DOI: 10.1098/rspa.2017.0655. 117

[11] Lekner, J. 2018b. *Theory of Electromagnetic Pulses*, IOP Concise Physics, Morgan & Claypool, San Rafael, CA. DOI: 10.1088/978-1-6432-7022-7. 117

[12] Lekner, J. 2020. Comparison of electromagnetic beams, *Optics Communications*, 458(124844):8. DOI: 10.1016/j.optcom.2019.124844. 122

[13] Torres, J. P. and Torner, L. (Eds.) 2011. *Twisted Photons*, Wiley-VCH. DOI: 10.1002/9783527635368. 112

6.5 ADDITIONAL REFERENCES (CHRONOLOGICAL ORDER)

[14] Barnett, S. M. and Allen, L. 1994. Orbital angular momentum and nonparaxial light beams, *Optics Communications*, 110:670–678 DOI: 10.1887/0750309016/b1142c14.

[15] Porras, M. A. 1994. The best quality optical beam beyond the paraxial approximation, *Optics Communications*, 111:338–349. DOI: 10.1016/0030-4018(94)90475-8.

[16] Berry, M. V. 1998. Wave dislocation reactions in non-paraxial Gaussian beams, *Journal of Modern Optics*, 45:1845–1858. DOI: 10.1080/09500349808231706.

[17] Nye, J. F. 1998. Unfolding of higher-order wave dislocations, *Journal of the Optical Society of America*, 15:1132–1138 DOI: 10.1364/josaa.15.001132.

[18] Zamboni-Rached, M. 2009. Unidirectional decomposition method for obtaining exact localized wave solutions totally free of backward components, *Physics Review A*, 79(013816). DOI: 10.1103/physreva.79.013816.

[19] Garay-Avendaño, R. L. and Zamboni-Rached, M. 2014. Exact analytic solutions of Maxwell's equations describing nonparaxial electromagnetic beams, *Applied Optics*, 53:4524–4531. DOI: 10.1364/AO.53.004524.

[20] Zamboni-Rached, M., Ambrosio, L. A., Dorrah, A. H., and Mojahedi, M. 2017. Structuring light under different polarization states within micrometer domains: Exact analysis from the Maxwell equations, *Optics Express*, 25:10051–10056. DOI: 10.1364/oe.25.010051.

[21] Philbin, T. G. 2018. Some exact solutions for light beams, *Journal of Optics*, 20(105603). DOI: 10.1088/2040-8986/aade6d.

[22] Lekner, J. and Andrejic, P. 2018. Nonexistence of exact solutions agreeing with the Gaussian beam on the beam axis or in the focal plane, *Optics Communications*, 407:22–26 DOI: 10.1016/j.optcom.2017.08.071.

CHAPTER 7

Sound Beams and Particle Beams

Sound beams and particle beams have properties in common with light beams, but also interesting differences. This Chapter is in two separate parts, each with its own set of references.

7.1   PART A: ACOUSTIC BEAMS

In Part A of this chapter c stands for the speed of sound, p, ρ, and v are the local pressure, density, and fluid velocity (all functions of the position r and the time t). The pressure, density, and fluid velocity in the undisturbed fluid are p_0, ρ_0 and v_0. The first two are constant in space and time, the latter is conventionally taken to be zero: we work in the rest frame if the fluid in which the sound propagates has a uniform flow. Linear sound is represented by solutions of the homogeneous wave equation (Landau and Lifshitz 1959 [3])

$$\left(\nabla^2 - \partial_{ct}^2\right) \Psi\left(r,t\right) = 0. \tag{7A.1}$$

Here Ψ is a complex velocity potential. The real first-order fluid velocity is given by the gradient of either the real or the imaginary part of Ψ:

$$v_1 = \nabla V, \quad V = Re\,\Psi \quad \text{or} \quad V = Im\,\Psi. \tag{7A.2}$$

The corresponding first-order density ρ_1 and pressure p_1 are then given by

$$\rho_1 = -\frac{\rho_0}{c^2}\partial_t V, \quad p_1 = c^2\rho_1 = -\rho_0\partial_t V. \tag{7A.3}$$

It has been shown (Lekner 2006a [5]) that the second-order quantities are also determined by Ψ, because they satisfy inhomogeneous linear second-order partial differential equations where the source terms are second-order in Ψ.

We shall consider single-frequency solutions of the wave equation (7A.1). We write $\Psi\left(r,t\right) = e^{-i\omega t}\psi\left(r\right)$ Then $\psi\left(r\right)$ satisfies the Helmholtz equation (1.5)

$$(\nabla^2 + k^2)\psi = 0, \quad k = \omega/c. \tag{7A.4}$$

The wave equation and the Helmholtz equation are separable in cylindrical polar coordinates $(r,\ \phi, z)$, where $r = \sqrt{x^2 + y^2}$ is the distance from the z-axis (we shall use r instead of ρ in

Part A because ρ is in use as the fluid density). A general solution which is non-singular on the z-axis (i.e., at $r = 0$) can be written as

$$\psi_m(r) = e^{im\phi} \int_0^k d\kappa \; f(k,\kappa) e^{iqz} J_m(\kappa r), \quad \kappa^2 + q^2 = k^2. \tag{7A.5}$$

Note that the range of κ ensures that both κ and q are real, so there are no terms with exponentially increasing or decreasing amplitude contributing to ψ_m. The function $f(k,\kappa)$ can be complex. We shall restrict ourselves to those $f(k,\kappa)$ for which the following integrals exist:

$$\int_0^k d\kappa \; \sqrt{\kappa} \, f(k,\kappa), \quad \int_0^k d\kappa \; \kappa^{-1} |f(k,\kappa)|^2, \quad \int_0^k d\kappa \; \kappa^{-1} q |f(k,\kappa)|^2. \tag{7A.6}$$

The existence of the first of these enables us to use Hankel's integral formula (Watson 1944 [11], Section 14.4; Lekner 2004 [4])

$$\int_0^\infty dr \; r \, J_m(\kappa r) J_m(\kappa' r) = \kappa^{-1} \delta(\kappa - \kappa'). \tag{7A.7}$$

The last two integrals give the energy E' and momentum P_z' contents per unit length of a generalized Bessel acoustic beam, as we shall see. The angular momentum is also given by the second integral: we shall prove that $ck J_z' = m E'$.

The energy, momentum and angular momentum densities of a beam

The energy density in a fluid is (Landau and Lifshitz 1959 [3], Lekner 2006a [5])

$$e(r,t) = e_0 + e_1(r,t) + e_2(r,t) + \dots \tag{7A.8}$$

We note that (r,t) is a point in space-time, not the position of a fluid particle: the Eulerian rather than the Lagrangian formulation is being used. The zero-order term e_0 is not associated with the sound wave. The first-order term is $e_1 = -(e_0 + p_0)c^{-1}\partial_{ct}V$, and cycle-averages to zero when V is linear in $\cos \omega t$ and $\sin \omega t$. The second-order term is

$$e_2(r,t) = \frac{1}{2}\rho_0 \left[(\nabla V)^2 + (\partial_{ct} V)^2 \right]. \tag{7A.9}$$

When V is equal to either the real or the imaginary part of $\Psi(r,t) = e^{-i\omega t}\psi(r)$, the cycle-average of e_2 is

$$\bar{e}(r) = \frac{1}{4}\rho_0 \left[|\nabla \psi|^2 + k^2 |\psi|^2 \right]. \tag{7A.10}$$

The momentum density is $p = \rho v$, where $\rho = \rho_0 + \rho_1 + \rho_2 + \dots$ and $v = v_1 + v_2 + \dots$. The first-order term $\rho_0 v_1$ cycle-averages to zero. There are two second-order terms, $\rho_1 v_1$ and $\rho_0 v_2$. The latter term is omitted by Landau and Lifshitz (1959) [3], but since v_2 is irrotational

(Lekner 2006a [5]) it can be expressed as the gradient of a potential, $v_2 = \nabla W$. From Equations (A6) and (A7) of Lekner (2006a) [5] we find that W satisfies the inhomogeneous wave equation

$$c^2 \nabla^2 W - \partial_t^2 W = 2(\partial_t \nabla V) \cdot \nabla V + 2\alpha(\partial_t V)\nabla^2 V, \qquad (7A.11)$$

where the constant α is given there by Equation (8). Cycle averaging of (7A.11), that is operating with $T^{-1} \int_0^T dt$, where $T = 2\pi/\omega$ is the time period everywhere in the beam, gives zero on the right-hand side. To show this, we write the velocity potential as

$$V(\mathbf{r},t) = C(\mathbf{r})\cos\omega t + S(\mathbf{r})\sin\omega t. \qquad (7A.12)$$

Denoting cycle average by a bar, we have

$$\overline{(\partial_t \nabla V) \cdot \nabla V} = -\omega\, \nabla C \cdot \nabla S\, \overline{(\cos^2\omega t - \sin^2\omega t)} = 0$$
$$\overline{(\partial_t V)\nabla^2 V} = -\omega\, k^2 C S\, \overline{(\cos^2\omega t - \sin^2\omega t)} = 0. \qquad (7A.13)$$

(We have used the fact that C and S satisfy the Helmholtz equation.) Also $\overline{\partial_t^2 W} = 0$. Thus $\nabla^2 \bar{W} = 0$: $\overline{W(\mathbf{r})}$ is a harmonic function, which cannot have a maximum or a minimum except at domain boundaries. For acoustic beams in unbounded media this implies that $\bar{W}$ is constant in space. Thus $\overline{v_2}$ is zero.

The non-zero cycle-averaged part of the momentum density is thus

$$\overline{\mathbf{p}(\mathbf{r})} = \overline{\rho_1 \mathbf{v}_1} = -\frac{\rho_0}{c}\overline{(\partial_{ct} V)\nabla V} = \frac{k\rho_0}{2c}Im(\psi^* \nabla \psi). \qquad (7A.14)$$

(This holds for both choices of the velocity potential, $V = Re\Psi$ and $V = Im\Psi$.) The result (7A.14) is in close correspondence with the momentum density in quantum mechanics (Lekner 2004 [4])

$$\mathbf{p}(\mathbf{r},t) = \hbar\, Im(\Psi^* \nabla \Psi), \qquad (7A.15)$$

where the wavefunction Ψ is a solution of the time-dependent Schrödinger equation. Quantum particle beams are discussed in Part B (Section 7.3).

We have chosen the acoustic beam propagation direction to be along the z-axis. The momentum component along the beam axis has the cycle average $\overline{p_z}$. We have, from (7A.10) and (7A.14),

$$\bar{e} - c\overline{p_z} = \frac{\rho_0}{4}|(\nabla - ik\hat{z})\psi|^2 \geq 0. \qquad (7A.16)$$

The cycle-average of the energy density is never smaller than c times the cycle-average of the momentum density component along the beam propagation direction. Equality instead of inequality in (7A.16) is possible only for a transversely infinite plane wave beam, $\psi \sim e^{ikz}$. The quantities on the left of (7A.16) are the Eulerian cycle-averaged energy density and momentum density. The local Eulerian cycle-averaged energy *flow* and momentum density are equal (Lighthill 1978 [10]).

Finally, we need the z-component of the cycle-averaged angular momentum density. We have, omitting the $\rho_0 v_2$ term which has zero cycle-average,

$$j_z = (\mathbf{r} \times \mathbf{p})_z = x p_y - y p_x = r p_\phi = -\left(\frac{\rho_0}{c}\right)(\partial_{ct} V)(\partial_\phi V). \tag{7A.17}$$

Note that j_z is zero unless the velocity potential has azimuthal dependence. The cycle-average of (7A.17) is, for $V = Re\,\psi$ or $Im\,\psi$,

$$\overline{j_z} = \frac{k\rho_0}{2c} Im(\psi^* \partial_\phi \psi). \tag{7A.18}$$

We take the azimuthal dependence to be entirely in the factor $e^{im\phi}$, as in (7A.5). Then

$$\overline{j_z} = \frac{k\rho_0}{2c} m |\psi|^2. \tag{7A.19}$$

The cycle-averaged angular momentum density is proportional to m.

Energy, momentum and angular momentum per unit length of beam

We shall calculate the cycle-averaged contents per unit length of the beam, specifically

$$E' = \int d^2r\, \bar{e}, \quad P_z' = \int d^2r\, \overline{p_z}, \quad J_z' = \int d^2r\, \overline{j_z}. \tag{7A.20}$$

Here $\int d^2r$ stands for $\int_{-\infty}^{\infty} dx \int_{-\infty}^{\infty} dy = \int_0^\infty dr\, r \int_0^{2\pi} d\phi$: we are taking a transverse slice through the beam. The reason for the prime notation is as follows: $dE = E'dz$ is the energy content in a slice of thickness dz of the beam, so E' may be viewed as dE/dz, et cetera.

Let us calculate J_z' first, because it is the simplest. We have, from (7A.5), (7A.19), and (7A.20),

$$J_z' = 2\pi \frac{k\rho_0}{2c} m \int_0^\infty dr\, r \int_0^k d\kappa\, f^*(\kappa) e^{-iqz} J_m(\kappa r) \int_0^k d\kappa'\, f(\kappa') e^{iq'z} J_m(\kappa'r)$$

$$= \pi \frac{k\rho_0}{c} m \int_0^k d\kappa\, \kappa^{-1} |f(\kappa)|^2. \tag{7A.21}$$

The last expression follows from the Hankel integral formula (7A.7).

Next, consider the momentum content per unit length. This is

$$P_z' = 2\pi \frac{k\rho_0}{2c} Im \int_0^\infty dr\, r \int_0^k d\kappa\, f^*(\kappa) e^{-iqz} J_m(\kappa r) \int_0^k d\kappa'\, f(\kappa') iq'\, e^{iq'z} J_m(\kappa'r)$$

$$= \pi \frac{k\rho_0}{c} \int_0^k d\kappa\, \kappa^{-1} q |f(\kappa)|^2. \tag{7A.22}$$

Finally, the energy content per unit length. We have

$$|\nabla \psi|^2 = (\nabla \psi^*) \cdot \nabla \psi = |\partial_r \psi|^2 + |\partial_z \psi|^2 + r^{-2} |\partial_\phi \psi|^2$$

$$= |\partial_r \psi|^2 + |\partial_z \psi|^2 + \frac{m^2}{r^2} |\psi|^2. \tag{7A.23}$$

(The last line applies only to wavefunctions with $e^{im\phi}$ azimuthal dependence.) In integrating over $|\partial_r \psi|^2$ we have the r-integration

$$\int_0^\infty dr\, r\, [\partial_r J_m(kr)]\, [\partial_r J_m(k'r)] = -\int_0^\infty dr\, r \frac{1}{r} \partial_r [r \partial_r J_m(kr)]\, J_m(k'r)$$

$$= \int_0^\infty dr\, r \left(k^2 - \frac{m^2}{r^2} \right) J_m(kr) J_m(k'r). \tag{7A.24}$$

The last equality follows because $J_m(kr)$ satisfies the differential equation

$$\frac{1}{r} \partial_r [r \partial_r J_m(kr)] + \left(k^2 - \frac{m^2}{r^2} \right) J_m(kr) = 0. \tag{7A.25}$$

Thus the m^2/r^2 terms cancel, and we are left with

$$E' = 2\pi \frac{\rho_0}{4} \int_0^\infty dr\, r \int_0^k d\kappa\, f^*(\kappa) e^{-iqz} J_m(\kappa r) \int_0^k d\kappa'\, f(\kappa') e^{iq'z} J_m(\kappa' r) \{\kappa^2 + qq' + k^2\}$$

$$= \pi \rho_0 k^2 \int_0^k d\kappa\, \kappa^{-1} |f(\kappa)|^2. \tag{7A.26}$$

From (7A.21), (7A.22), and (7A.26) we see that

$$cP'_z \leq E', \qquad ckJ'_z = mE'. \tag{7A.27}$$

The inequality follows from the fact that $q = \sqrt{k^2 - \kappa^2} \leq k$ within the range of integration. The equality can be written as (see also Hefner and Marston 1998, 1999 [1, 2])

$$\omega J'_z = mE', \tag{7A.28}$$

and is in accord with the sound beam being made up of phonons, each of energy $\hbar\omega$ and each carrying angular momentum $\hbar m$. We may, if we wish, think of each phonon carrying momentum $\hbar k$ (with $k = \omega/c$). This is consistent with the inequality in (7A.27) because not all the phonons are travelling parallel to the z-axis: a transversely finite acoustic beam is necessarily either converging or diverging to some extent. More precisely, each phonon carries the z component of momentum equal to $\hbar q$. A similar interpretation can be made for sound pulses (Lekner 2006c [7]).

The results we have just given are exact in the absence of scattering and viscous damping. They are summarized in the following:

$$
\begin{bmatrix} E' \\ c P'_z \\ c J'_z \end{bmatrix} = \pi \rho_0 k \int_0^k d\kappa \, \kappa^{-1} |f(k,\kappa)|^2 \begin{bmatrix} k \\ q \\ m \end{bmatrix}. \tag{7A.29}
$$

Similar expressions have been derived for the total energy, momentum and angular momentum of sound pulses (Lekner 2017 [9]). Acoustic pulses may also be characterized by a wavenumber weight function, which in the case of pulses is integrated over the total wavenumber k as well as over the transverse wavenumber component κ:

$$
\begin{bmatrix} E \\ c P_z \\ c J_z \end{bmatrix} = \pi^2 \rho_0 \int_0^\infty dk \int_0^k d\kappa \, \kappa^{-1} q \, |f(k,\kappa)|^2 \begin{bmatrix} k \\ q \\ m \end{bmatrix}. \tag{7A.30}
$$

For the beams ψ_m of Section 2.8 derived from ψ_0 the ratio $c P'_z / E'$ decreases with m: for $m = 0$, 1, 2 it is $2/3$, $8/15$, $16/35$, respectively. The expression for general m is

$$
\frac{c P'_z}{E'} = \frac{2^{m+1} (m+1)!}{(2m+3)!!} = \frac{1}{2}\sqrt{\frac{\pi}{m}} + O\left(m^{-\frac{3}{2}}\right). \tag{7A.31}
$$

Figure 7.1 shows the focal plane variation of $\bar{e}$ and $c\overline{p_z}$ of an acoustic beam based on the proto-beam ψ_0. Both are maximal at the origin. The radial and azimuthal components of momentum density are zero in the focal plane $z = 0$. The longitudinal component is

$$
c\overline{p_z}(r,0) = k\rho_0 (kr)^{-4} J_1(kr)[\sin kr - kr \cos kr] = k\rho_0 (kr)^{-2} J_1(kr) j_1(kr). \tag{7A.32}
$$

The first zero of $c\overline{p_z}$ in the focal plane occurs at the first zero of J_1, at $kr \approx 3.8317$. The next zero is that of $\sin kr - kr \cos kr$, at 4.4934. The energy density is non-negative, and decays asymptotically as r^{-3}. The energy and momentum per unit length of beam are (when ψ_0 is dimensionless and normalized to unity at the origin) $E' = 2\pi\rho_0$ and $c P'_z = \frac{4}{3}\pi\rho_0$, with ratio $\frac{2}{3}$. These values are in agreement with (7A.29) and (7A.31). The energy density is never smaller than c times the momentum density, in accord with (7A.16).

Figure 7.2 gives the longitudinal view of the sound beam based on the proto-beam ψ_0. On axis the energy density and momentum density take the forms (we set $kz = \zeta$)

$$
\bar{e}(0,z) = \frac{2}{\zeta^6}\{\zeta^4 + \zeta^2 + 4 + (\zeta^2 - 4)\cos\zeta - \zeta(\zeta^2 + 4)\sin\zeta\}
$$

$$
c\overline{p_z}(0,z) = \frac{2}{\zeta^3}\{\zeta - \sin\zeta\}. \tag{7A.33}
$$

Both are maximal at $z = 0$, and both have the asymptotic value $2\zeta^{-2}$ at large $k|z|$.

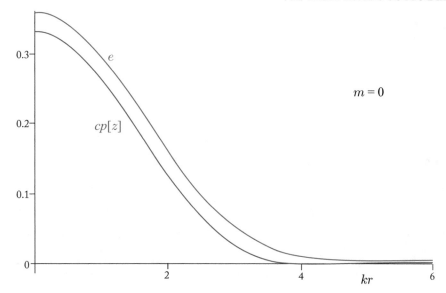

Figure 7.1: The focal plane variation of $\bar{e}$ and $c\,\overline{p_z}$ of an acoustic beam based on the proto-beam wavefunction ψ_0. Only the longitudinal component of momentum density is non-zero in the focal plane.

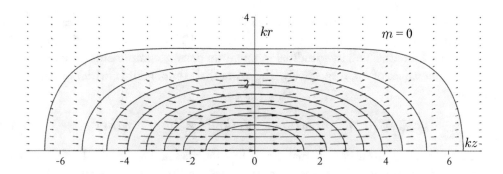

Figure 7.2: Longitudinal section of the acoustic beam based on ψ_0. The energy density is shown in contours and shading, the momentum density by arrows. The azimuthal component of the momentum density is everywhere zero.

Finally, we shall look at the properties of the $m = 1$ sound beam constructed from $\psi_1 = -k^{-1}e^{i\phi}\partial_\rho\psi_0$ of Section 2.8. It has the wavenumber weight function $f_1(k,\kappa) = 2\kappa^2/k^3$. The energy and momentum per unit length of beam are (when ψ_0 is dimensionless and normalized to unity at the origin) $E' = \pi\rho_0$ and $cP'_z = \frac{8}{15}\pi\rho_0$, with ratio $\frac{8}{15}$. The angular momentum per unit length is given by $ck\,J'_z = E'$. These values are calculated by direct integration, or from (7A.29).

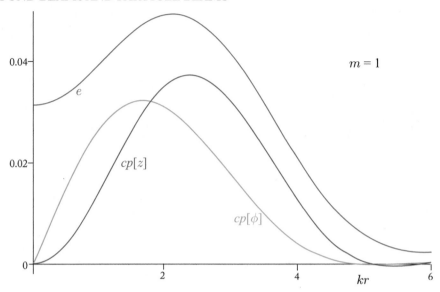

Figure 7.3: The focal plane variation of $\bar{e}$, $c\overline{p_\phi}$ and $c\overline{p_z}$ of an acoustic beam based on the proto-beam wavefunction ψ_1. The radial component of the momentum density is zero in the focal plane.

Figures 7.3 and 7.4 show the energy and momentum densities in the focal plane. Note that the energy density is maximum on the circle $k\rho \approx 2.1292$, rather than at the center of the focal plane. The azimuthal and longitudinal momentum density components peak at somewhat lower and higher values of $k\rho$, respectively. The azimuthal component is, with normalization as stated above,

$$c\overline{p_\phi} = 2(kr)^{-5}\{krJ_0(kr) - 2J_1(kr)\}^2 = 2(kr)^{-3}J_2(kr)^2. \tag{7A.34}$$

It is non-negative, and linear in r at small kr. It first touches zero at $kr \approx 5.1356$. The angular momentum density $j_z = rp_\phi$ is thus also non-negative, and rises quadratically with r at small kr.

Figure 7.5 gives the energy and momentum densities of the sound beam based on ψ_1, in longitudinal section. The azimuthal component of momentum is not shown.

Part A (Section 7.1) is based mainly on Lekner 2006b [6], with new material introduced in the examples. Tensor invariants derived from the conservation of momentum and of angular momentum are discussed in Lekner 2007 [8].

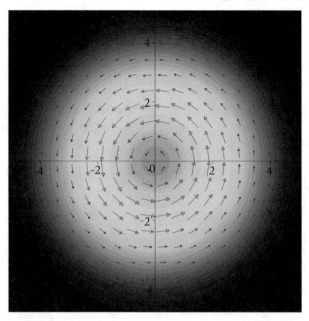

Figure 7.4: A density plot of the energy density (light color), combined with a field-plot of the azimuthal momentum density (arrows). The plot extends over $k\,|x| \leq 5$ horizontally, and $k\,|y| \leq 5$ vertically. Drawn in the focal plane $z = 0$ for an acoustic beam based on the beam wavefunction ψ_1.

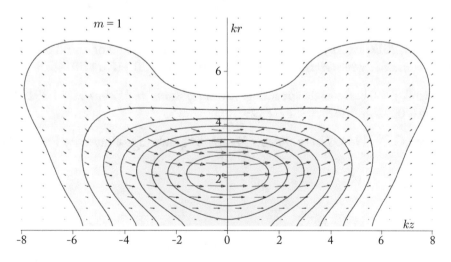

Figure 7.5: Longitudinal section of the acoustic beam based on ψ_1. The energy density is shown in contours and shading, the momentum longitudinal and radial density components by arrows. The azimuthal component of the momentum density is transverse to the figure.

7.2   REFERENCES FOR PART A

[1] Hefner, B. T. and Marston, P. L. 1998. Acoustical helicoidal waves and Laguerre-Gaussian beams: Applications to scattering and to angular momentum transport, *Journal of the Acoustical Society of America*, 103(2971). DOI: 10.1121/1.422390. 131

[2] Hefner, B. T. and Marston, P. L. 1999. An acoustical helicoidal wave transducer with applications for the alignment of ultrasonic and underwater systems, *Journal of the Acoustical Society of America*, 106:3313–3316. DOI: 10.1121/1.428184. 131

[3] Landau, L. D. and Lifshitz, E. M. 1959. *Fluid Mechanics*, Chapter VIII, Pergamon, Oxford. 127, 128

[4] Lekner, J. 2004. Invariants of atom beams, *Journal of Physics B: Atomic and Molecular Optic Physics*, 37:1725–1736. DOI: 10.1088/0953-4075/37/8/013. 128, 129

[5] Lekner, J. 2006a. Energy and momentum of sound pulses, *Physica A*, 363:217–225. DOI: 10.1016/j.physa.2005.08.045. 127, 128, 129

[6] Lekner, J. 2006b. Acoustic beams with angular momentum, *Journal of the Acoustical Society of America*, 120:3475–3478. DOI: 10.1121/1.2360420. 134

[7] Lekner, J. 2006c. Angular momentum of sound pulses, *Journal of Physics: Condensed Matter*, 18:6149–6158 DOI: 10.1088/0953-8984/18/26/032. 131

[8] Lekner, J. 2007. Acoustic beam invariants, *Physics Review E*, 57(036610). DOI: 10.1103/physreve.75.036610. 134

[9] Lekner, J. 2017. Energy, momentum, and angular momentum of sound pulses, *Journal of the Acoustical Society of America*, 142:3428–3435. DOI: 10.1121/1.5014058. 132

[10] Lighthill, J. 1978. Acoustic streaming, *Journal of Sound and Vibration*, 61:391–418. DOI: 10.1016/0022-460x(78)90388-7. 129

[11] Watson, G. N. 1944. *Theory of Bessel Functions*, Cambridge University Press. 128

7.3 PART B: PARTICLE BEAMS

We shall consider the simplest case of quantum particle beams, namely coherent beams of spinless particles. The collective wavefunction $\Psi(r, t)$ is assumed to represent the probability amplitude of the beam. Further, it is assumed that $\Psi(r, t)$ evolves in time according to Schrödinger's time-dependent equation $H\Psi(r, t) = i\hbar\partial_t\Psi(r, t)$, where H is an effective Hamiltonian, the same form as for one atom in some external potential $V(r)$:

$$H = -\frac{\hbar^2}{2M}\nabla^2 + V(r).\tag{7B.1}$$

The equation of continuity follows from Schrödinger's time-dependent equation:

$$\partial_t|\Psi|^2 + \nabla \cdot \left\{\frac{\hbar}{M}Im\,(\Psi^*\nabla\Psi\right\} = 0.\tag{7B.2}$$

The same form of continuity equation (conservation of probability) holds in the many-particle case (Lekner 2004a [13]).

A real momentum density may be defined in terms of the operator $\hat{p} = -i\hbar\nabla$:

$$p(r, t) = \frac{1}{2}\left\{\Psi^*\hat{p}\Psi + \Psi(\hat{p}\Psi)^*\right\} = \frac{i\hbar}{2}\left\{\Psi\nabla\Psi^* - \Psi^*\nabla\Psi\right\} = \hbar\,Im(\Psi^*\nabla\Psi).\tag{7B.3}$$

Hence the continuity Equation 7B.2 reads

$$\partial_t|\Psi|^2 + \frac{1}{M}\nabla \cdot p = 0.\tag{7B.4}$$

In the case of a 'steady' beam in which Ψ is an energy eigenstate, $H\Psi = \varepsilon\Psi$, the time variation has the form $\Psi(r, t) = e^{-\frac{i\varepsilon t}{\hbar}}\psi(r)$, the probability density $|\Psi|^2 = |\psi|^2$ is independent of time, and hence $\nabla \cdot p = 0$. (For beams which oscillate in time we can average over the oscillations, denote such averages by a bar, and obtain $\nabla \cdot \bar{p} = 0$; such extensions will be understood throughout this section.)

The first beam invariant follows from operating on $\nabla \cdot p = 0$ (or $\nabla \cdot \bar{p} = 0$) with $\int d^2r = \int_{-\infty}^{\infty} dx \int_{-\infty}^{\infty} dy = \int_0^{\infty} d\rho\,\rho \int_0^{2\pi} d\phi$ (we are using cylindrical coordinates ρ, ϕ, z). The terms $\partial_x p_x$ and $\partial_y p_y$ integrate to zero when the beam is propagating in the z direction and is finite transversely, since for example p_x goes to zero at $x = \pm\infty$, and $\int d^2r\,\partial_x p_x = \int_{-\infty}^{\infty} dy \int_{-\infty}^{\infty} dx \partial_x p_x = 0$. What remains is

$$\partial_z \int d^2r\,p_z = 0, \quad \text{or} \quad P_z' = \int d^2r\,p_z = \text{constant}.\tag{7B.5}$$

The quantity P_z', constant along the beam extent, may be viewed as dP_z/dz, since $dP_z = P_z'dz$ is the momentum content in a slice of thickness dz of the beam. The momentum content per unit length of the beam is the same everywhere along the length of the beam: P'_z is an invariant. Note that this invariance follows not from momentum conservation but from the continuity equation; it comes from the proportionality of the probability density flux to the momentum density.

Conservation of momentum, and of angular momentum

Momentum conservation for the electromagnetic field is expressed in terms of the Maxwell tensor. Lekner (2004a) [13] wrote down an analogous momentum conservation law for a particle field described by a wavefunction $\Psi(r,t)$. From the definition $p(r,t) = \hbar Im(\Psi^* \nabla \Psi)$ in (7B.3) and Schrödinger's time evolution equation $H\Psi(r,t) = i\hbar \partial_t \Psi(r,t)$ it follows that

$$\partial_t p = \frac{1}{2}\left\{(H\Psi)\nabla\Psi^* - \Psi\nabla(H\Psi^*) + (H\Psi^*)\nabla\Psi - \Psi^*\nabla(H\Psi)\right\}. \qquad (7B.6)$$

If the Hamiltonian is of the form (7B.1) the momentum conservation law reads (Lekner 2004a [13])

$$\partial_t p_i + \sum_j \partial_j \tau_{ij} = f_i \qquad (i,j = x,y,z). \qquad (7B.7)$$

The potential part V of the Hamiltonian (2B.1) gives the force density $f(r,t) = (-\nabla V)\Psi^*\Psi$. The kinetic part gives the real and symmetric momentum flux density tensor

$$\tau_{ij} = \frac{\hbar^2}{2M}\left\{(\partial_i\Psi^*)\partial_j\Psi + (\partial_i\Psi)\partial_j\Psi^* - \frac{1}{2}\partial_i\partial_j\Psi^*\Psi\right\}. \qquad (7B.8)$$

Note that it is different from the 'stress tensor of the probability fluid' (Madelung 1926, Bialynicki-Birula, Cieplak and Kaminski 1992, p.89 [12, 15]),

$$T_{ij} = \frac{\hbar^2}{4M}\Psi^*\Psi\partial_i\partial_j\ln\Psi^*\Psi. \qquad (7B.9)$$

Lekner (2004a) [13] shows that three beam invariants follow from the momentum conservation law (7B.7), in the absence of external force.

For a beam of spinless particles the angular momentum density is given by

$$j(r,t) = r \times p(r,t), \quad \text{or} \quad j_i = \sum_{j,k} \varepsilon_{ijk} r_j p_k. \qquad (7B.10)$$

The angular momentum flux density tensor is defined in terms of the momentum flux density tensor:

$$\mu_{li} = \sum_{j,k} \varepsilon_{ijk} r_j \tau_{kl}. \qquad (7B.11)$$

In the absence of external torques the angular momentum density satisfies the conservation law

$$\partial_t j_i + \sum_l \partial_l \mu_{li} = 0. \qquad (7B.12)$$

Application of $\int d^2r$ the angular momentum conservation law gives three more beam invariants (Lekner 2004a, Section 4 [13]). Thus seven beam invariants follow from the conservation of probability (the equation of continuity), momentum and angular momentum.

Generalized Bessel beams

For generalized Bessel beams (those based on wavefunctions of the form (7B.14)) there are two more quantities of physical import which are constant along the length of the beam: the energy E' and angular momentum content J_z' per unit length of a beam in free space. We shall calculate the cycle-averaged contents per unit length of the beam, specifically

$$E' = \int d^2r \, \bar{e}, \quad P_z' = \int d^2r \, \overline{p_z}, \quad J_z' = \int d^2r \, \overline{j_z}. \tag{7B.13}$$

As above, $\int d^2r$ stands for $\int_{-\infty}^{\infty} dx \int_{-\infty}^{\infty} dy = \int_0^\infty d\rho \, \rho \int_0^{2\pi} d\phi$. Because $dE = E'dz$ is the energy content in a slice of thickness dz of the beam, E' may be viewed as dE/dz, et cetera.

We shall derive expressions for the quantities in (7B.13) for particle beams with wavefunctions of the form

$$\psi_m(\rho, \phi, z) = e^{im\phi} \int_0^k d\kappa \, f(k, \kappa) e^{iqz} J_m(\kappa\rho), \qquad q = \sqrt{k^2 - \kappa^2}. \tag{7B.14}$$

In the absence of external forces, the Hamiltonian is just the kinetic energy operator

$$\frac{-\hbar^2}{2M} \nabla^2 = \frac{-\hbar^2}{2M} \left(\partial_\rho^2 + \rho^{-1} \partial_\rho + \rho^{-2} \partial_\phi^2 + \partial_z^2 \right). \tag{7B.15}$$

For generalized Bessel beams with wavefunctions (7B.14),

$$\left(\partial_\rho^2 + \rho^{-1} \partial_\rho + \rho^{-2} \partial_\phi^2 + \partial_z^2 \right) e^{im\phi} e^{iqz} J_m(\kappa\rho) = \left(-\kappa^2 - q^2 \right) e^{im\phi} e^{iqz} J_m(\kappa\rho). \tag{7B.16}$$

Hence the kinetic energy density is

$$\begin{aligned} e(\boldsymbol{r}) &= \frac{\hbar^2 k^2}{2M} |\psi_m|^2 \\ &= \frac{\hbar^2 k^2}{2M} \int_0^k d\kappa \, f^*(k, \kappa) e^{-iqz} J_m(\kappa\rho) \int_0^k d\kappa' \, f(k, \kappa') e^{iq'z} J_m(\kappa'\rho). \end{aligned} \tag{7B.17}$$

The primed variables satisfy $q' = \sqrt{k^2 - \kappa'^2}$. The resulting energy content per unit length of the beam is $E' = 2\pi \int_0^\infty d\rho \, \rho e(\boldsymbol{r})$, in which we shall use the Hankel formula (Watson 1944, Section 14.4 [16]; see also Lekner 2004b, Appendix A [14])

$$\int_0^\infty d\rho \, \rho \, J_m(\kappa\rho) J_m(\kappa'\rho) = \kappa^{-1} \delta(\kappa - \kappa'). \tag{7B.18}$$

Thus

$$E' = \frac{\hbar^2 k^2}{2M} \, 2\pi \int_0^k d\kappa \, \kappa^{-1} |f(k,\kappa)|^2. \tag{7B.19}$$

In examining the energy content we have incidentally proved that the probability content per unit length of beam is a constant, and takes the value

$$N' = \int d^2r |\psi_m|^2 = 2\pi \int_0^k d\kappa \, \kappa^{-1} |f(k,\kappa)|^2. \tag{7B.20}$$

Next we look at the momentum content per unit length of beam. The momentum density is, from (7B.3), $\boldsymbol{p}(\boldsymbol{r}) = \hbar \, Im(\Psi^* \nabla \Psi)$. In cylindrical polar coordinates (ρ, ϕ, z) the gradient operator is $(\partial_\rho, \rho^{-1}\partial_\phi, \partial_z)$. We are interested in the longitudinal component

$$p_z = \hbar \, Im \left(\psi^* \partial_z \psi \right)$$

$$= \hbar \, Im \left\{ \int_0^k d\kappa \, f^*(k,\kappa) e^{-iqz} J_m(\kappa\rho) \int_0^k d\kappa' \, f(k,\kappa') iq' e^{iq'z} J_m(\kappa'\rho) \right\}. \tag{7B.21}$$

We integrate over a section of the beam and use (2B.18) to obtain

$$P'_z = \hbar 2\pi \int_0^k d\kappa \, \kappa^{-1} q \, |f(k,\kappa)|^2. \tag{7B.22}$$

Finally, we shall calculate J'_z. We have

$$j_z = x p_y - y p_x = \rho p_\phi = \hbar \, Im \left(\psi^* \partial_\phi \psi \right) = \hbar m \, \psi^* \psi. \tag{7B.23}$$

Hence the longitudinal component of the angular momentum content per unit length of the beam is, by the same calculation as for E',

$$J'_z = \int d^2r \, j_z = \hbar m 2\pi \int_0^k d\kappa \, \kappa^{-1} \, |f(k,\kappa)|^2 = \hbar m \left(2M/\hbar^2 k^2 \right) E'. \tag{7B.24}$$

The angular momentum and energy contents per unit length of the beam are proportional.

We summarize the results for generalized Bessel particle beams:

$$\begin{bmatrix} E' \\ P'_z \\ J'_z \end{bmatrix} = 2\pi \int_0^k d\kappa \, \kappa^{-1} \, |f(k,\kappa)|^2 \begin{bmatrix} \varepsilon_k \\ \hbar q \\ \hbar m \end{bmatrix}, \qquad \varepsilon_k = \hbar^2 k^2 / 2M. \tag{7B.25}$$

Such particle beams may therefore be regarded as superpositions, with probability amplitude $f(k,\kappa)$, of particle quanta with energies $\varepsilon_k = \hbar^2 k^2 / 2M$, longitudinal momentum component $\hbar q = \sqrt{2M\varepsilon_k - \hbar^2 \kappa^2}$, and longitudinal angular momentum component $\hbar m$. Tensor invariants

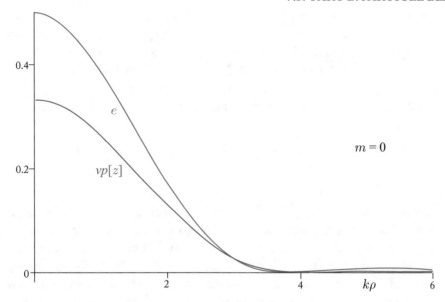

Figure 7.6: Variation of the energy density e and $v = \hbar k / 2M$ times p_z, in the focal plane of a particle beam based on the proto-beam wavefunction ψ_0 of Section 2.5.

arising from the conservation of momentum and of angular momentum are derived in Lekner 2004a [13].

In the figures that follow we plot energy density e and momentum densities times $v = \hbar k / 2M$. The velocity factor brings $v\boldsymbol{p}$ to the same dimension as e. Also, when a beam approximates the plane wave, for which $p_z \to \hbar k$, we have $v p_z \to \varepsilon_k = \hbar^2 k^2 / 2M$. Figure 7.6 shows the focal plane variation of the energy e and the velocity-momentum product $v p_z$ of a particle beam based on the proto-beam ψ_0 of Section 2.5, which we normalize to unity at the origin. Both e and p_z are maximal at the origin. In the graphs we set $\hbar = 1, M = 1, k = 1$. The energy density is then just $|\psi_0|^2 / 2$ from (7B.17), which is $2(k\rho)^{-2}[J_1 (k\rho)]^2$ in the focal plane. The radial and azimuthal components of momentum density are zero in the focal plane $z = 0$. The longitudinal component is

$$v p_z (\rho, 0) = 2(k\rho)^{-4} J_1 (k\rho) [\sin k\rho - k\rho \cos k\rho]$$
$$= 2(k\rho)^{-2} J_1(k\rho) j_1(k\rho). \tag{7B.26}$$

At the origin (the center of the focal region) we find, in the units specified above, $e(0,0) = 1/2$, $v p_z (0,0) = 1/3$. The first zero of p_z in the focal plane occurs at the first zero of $J_1(k\rho)$, at $k\rho \approx 3.8317$. This zero is shared by $e(\rho, 0)$. The next zero of p_z is that of $\sin k\rho - k\rho \cos k\rho$, at 4.4934. The energy density is non-negative, and decays asymptotically as ρ^{-3}. The momentum density has regions where it is slightly negative.

The energy and momentum per unit length of beam are (when ψ_0 is dimensionless and normalized to unity at the origin) $E' = 2\pi$ and $vP'_z = \frac{4}{3}\pi$, with ratio $\frac{2}{3}$. These values are in agreement with (7B.19) and (7B.22). The energy density may be smaller than v times the momentum density, in contrast to the relation $E' \geq cP'_z$ which holds for electromagnetic and acoustic beams (with c the speed of light and of sound, respectively).

On the beam axis the energy density and longitudinal component of momemntum take the forms (we set $kz = \zeta$)

$$e\,(0, z) = 2\zeta^{-4}[\zeta^2 - 2\zeta\sin\zeta - 2\cos\zeta + 2].$$

$$vp_z\,(0, z) = 2\zeta^{-3}[\zeta - \sin\zeta\,]. \tag{7B.27}$$

Both are asymptotic to $2\zeta^{-2}$. Figure 7.7 shows the ψ_0 particle beam in longitudinal section.

Next we shall look at the properties of the $m = 1$ particle beam constructed from $\psi_1 = -k^{-1}e^{i\phi}\partial_\rho\psi_0$ of Section 2.8. It has the wavenumber weight function $f_1\,(k, \kappa) = 2\kappa^2/k^3$. The energy and momentum per unit length of beam are (when ψ_0 is dimensionless and normalized to unity at the origin) $E' = \pi$ and $vP'_z = \frac{8}{15}\pi$, with ratio $\frac{8}{15}$. The angular momentum per unit length is given by $vkJ'_z = E'$. These values are calculated by direct integration, or from (7B.25).

Figures 7.8 and 7.9 show the energy and momentum densities in the focal plane. These densities all share the first non-axial zero on the circle $k\rho \approx 5.1356$, because they share the common factor $k\rho J_0\,(k\rho) - 2J_1\,(k\rho) = -k\rho J_2(k\rho)$. The energy density is zero at the center of the focal plane, and maximum on the circle $k\rho \approx 2.2999$. The azimuthal and longitudinal momentum density components peak at somewhat lower and higher values of $k\rho$, respectively. The azimuthal component is, with normalization as stated above,

$$vp_\phi\,(\rho, 0) = 2(k\rho)^{-5}\{k\rho J_0\,(k\rho) - 2J_1\,(k\rho)\}^2 = (k\rho)^{-1}e(\rho, 0). \tag{7B.28}$$

It is non-negative, and linear in ρ at small $k\rho$. It first touches zero at $k\rho \approx 5.1356$.

The angular momentum density $j_z = \rho p_\phi$ and the energy density are thus proportional. This is a general property of steady non-relativistic particle beams: the energy density is $e(r) = \varepsilon_k\psi^*\psi$ from (7B.17), and the angular momentum density is, from (7B.23),

$$j_z(r) = \rho p_\phi(r) = \hbar m\psi^*\psi = m\frac{2M}{\hbar k^2}e(r). \tag{7B.29}$$

We have already seen in (7B.25) that the energy and angular momentum contents per unit length of beam are proportional. The equality (7B.23) shows that the densities are proportional, everywhere in the beam.

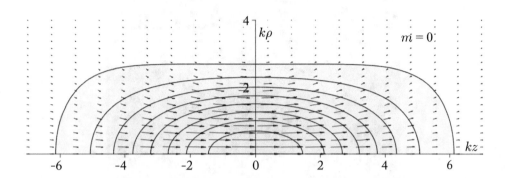

Figure 7.7: Longitudinal section of the particle beam based on ψ_0. The energy density is shown in contours and shading, the momentum density by arrows. The azimuthal component of the momentum density is everywhere zero.

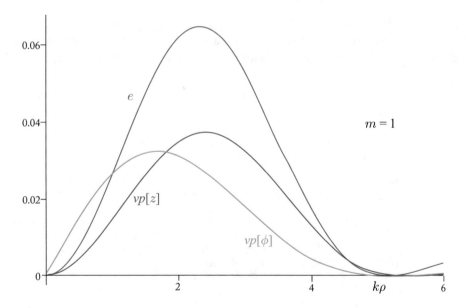

Figure 7.8: The focal plane variation of e, vp_ϕ and vp_z of a particle beam based on the proto-beam wavefunction ψ_1. The radial component of the momentum density is zero in the focal plane.

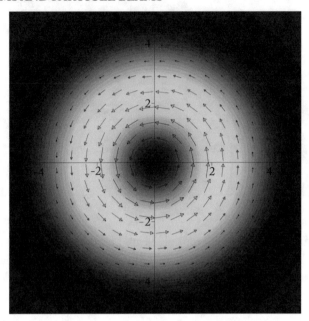

Figure 7.9: A density plot of the energy density (light color), combined with a field-plot of the azimuthal momentum density (arrows). The plot extends over $k\,|x| \leq 5$ horizontally, and $k\,|y| \leq 5$ vertically. Drawn in the focal plane $z = 0$ for the particle beam based on the wavefunction ψ_1.

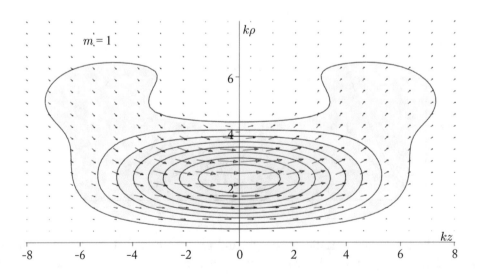

Figure 7.10: Longitudinal section of the particle beam based on ψ_1. The energy density is shown in contours and shading, the momentum longitudinal and radial density components by arrows. The azimuthal component of the momentum density is transverse to the figure.

7.4   REFERENCES FOR PART B

[12] Bialynicky-Birula, I., Cieplak, M., and Kaminsky, J. 1992. *Theory of Quanta*, p. 89, Oxford. 138

[13] Lekner, J. 2004a. Invariants of atom beams, *Journal of Physics B: Atomic and Molecular Optic Physics*, 37:1725–1736. DOI: 10.1088/0953-4075/37/8/013. 137, 138, 139, 141

[14] Lekner, J. 2004b. Invariants of three types of generalized Bessel beams, *Journal of Optics A: Pure Applied Optics*, 6:837–843. DOI: 10.1088/1464-4258/6/9/004. 139

[15] Madelung, E. 1926. Quantentheorie in hydrodynamischer form, *Zeitschrift für Physik*, 40:322–326. DOI: 10.1007/bf01400372. 138

[16] Watson, G. N. 1944. *Theory of Bessel Functions*, Cambridge University Press. 139

CHAPTER 8

Measures of Focal Extent

8.1 INTRODUCTION

All physical beams are localized transversely, and consequently converge onto or spread from a focal region. Of interest is the degree of localization. For any wave of frequency $\omega = ck$ we expect the minimum extent of a beam to be of order k^{-1}. We have given examples of possible longitudinal and transverse lengths as measures of focal extent in Section 2.7; for the proto-beam both were of order k^{-1}. Here we shall enlarge and generalize by introducing the Andrejic (2018a,b) [1, 2] measures, respectively a length for longitudinal extent of the focal region, and an area (with its associated radius) for the transverse extent. We discuss the degree of localization of two families of beams, each characterized by a dimensionless parameter kb, where b is a length. The longitudinal and transverse extents of the beams are different in their dependence on k and b: we shall show that at large kb these tend to πb and $\sqrt{b/k}$, respectively.

The beam families are those of Carter (1973) [4], and the one obtained from the proto-beam by a complex shift in z (Lekner 2016 [7]). Both are based on wavefunctions in the form (2.14) with $m = 0$:

$$\psi_C (\rho, z) = \frac{b/k}{1 - e^{-kb/2}} \int_0^k d\kappa\, \kappa\, e^{-b\kappa^2/2k + iqz} J_0 (\kappa\rho)$$

$$= \frac{b/k}{e^{kb/2} - 1} \int_0^k dq\, q\, e^{bq^2/2k + iqz} J_0 (\kappa\rho) \tag{8.1}$$

$$\psi_b (\rho, z) = \frac{b^2}{e^{kb} (kb - 1) + 1} \int_0^k d\kappa\, \kappa\, e^{qb + iqz} J_0 (\kappa\rho)$$

$$= \frac{b^2}{e^{kb} (kb - 1) + 1} \int_0^k dq\, q\, e^{qb + iqz} J_0 (\kappa\rho). \tag{8.2}$$

As the dimensionless parameter kb tends to zero, the families (8.1) and (8.2) both tend to the proto-beam (shown in Figure 8.1),

$$\psi_0 (\rho, z) = \frac{2}{k^2} \int_0^k d\kappa\, \kappa\, e^{iqz} J_0 (\kappa\rho) = \frac{2}{k^2} \int_0^k dq\, q\, e^{iqz} J_0 (\kappa\rho). \tag{8.3}$$

Figure 8.2 shows the ψ_C, ψ_b and proto beams in the focal plane, with ψ_C, ψ_b shown at $kb = 2$. The proto-beam first zero is at $k\rho \approx 3.83$ (the first zero of $J_1(k\rho)/k\rho$); at $kb = 2$ the first Carter zero is at $k\rho \approx 4.48$, followed by the first zero of ψ_b at $k\rho \approx 4.77$.

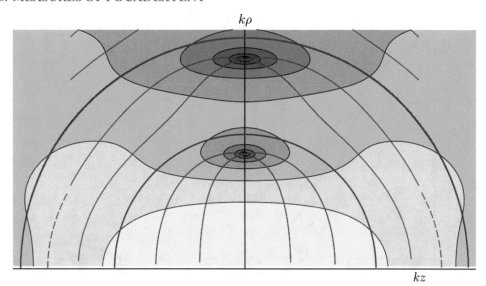

Figure 8.1: The proto-beam wavefunction $\psi_0\,(\rho, z)$ in the focal region: contours of constant modulus (logarithmic scale), and isophase surfaces (red and brown). The phase is chosen to be zero at the origin, and isophase contours are shown at intervals of $\pi/3$. The π, $-\pi$, and 2π, -2π phase surfaces (brown) are hemispherical. The isophase surfaces (apart from those which are integral multiples of π) meet on the zeros of ψ_0, which lie in the focal plane. The extent of the focal region shown is in $k\,|z| \leq 8$, $k\rho \leq 8$.

Figure 8.3 shows the Carter and ψ_b beams in the focal region, again at $kb = 2$. We note that at the same value of kb the Carter beam has slightly larger wavelength (as determined by the spacing between isophase contours). The contours of equal modulus are not at the same levels in the upper and lower parts of the figure.

Section 2.7 defined two lengths which, in general, characterize the extent of the focal region: the transverse localization length L_t, and the longitudinal localization length L_l. There are constraints on how we can define these lengths, because beam wavefunctions decay as a negative power of the distance from the focal region, times a sinusoidal function. For example, $\psi_0\,(\rho, 0) = 2(k\rho)^{-1} J_1(k\rho)$, asymptotically proportional to $\rho^{-3/2}$ times a sinusoidal function. The longitudinal and transverse extents of the focal region of a scalar beam are defined by

$$L_l = \left\langle |z|^{\frac{1}{2}} \right\rangle^2, \qquad L_t = \left\langle \rho^{\frac{1}{2}} \right\rangle^2 \tag{8.4}$$

$$\left\langle |z|^{\frac{1}{2}} \right\rangle = \frac{\int_0^\infty dz\, z^{\frac{1}{2}}\, |\psi(0, z)|^2}{\int_0^\infty dz\, |\psi(0, z)|^2}, \qquad \left\langle \rho^{\frac{1}{2}} \right\rangle = \frac{\int_0^\infty d\rho\, \rho\, \rho^{\frac{1}{2}}\, |\psi(\rho, 0)|^2}{\int_0^\infty d\rho\, \rho\, |\psi(\rho, 0)|^2}. \tag{8.5}$$

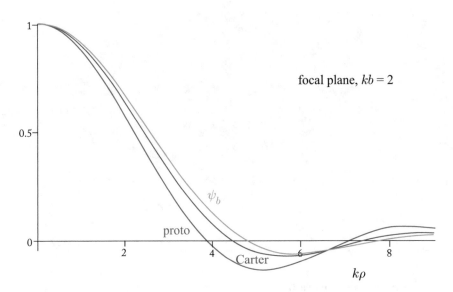

Figure 8.2: The Carter, ψ_b and proto beams in the focal plane, drawn in red, green and blue, respectively. ψ_C and ψ_b are shown for $kb = 2$. The proto-beam takes the form $2J_1(k\rho)/k\rho$ in the focal plane. It is the confluent form of ψ_C and ψ_b as $kb \to 0$.

For the fundamental Gaussian mode (2.7),

$$L_l = 2b, \qquad L_t = \Gamma^2\left(\frac{5}{4}\right)\left(\frac{b}{k}\right)^{\frac{1}{2}} \approx 0.82\left(\frac{b}{k}\right)^{\frac{1}{2}}. \tag{8.6}$$

Note that both lengths tend to zero as $b \to 0$. This is unphysical; the Gaussian beam is not an exact solution of the Helmholtz equation, and the discrepancy is worst in the long-wave limit $kb \to 0$.

For the proto-beam

$$L_l = \frac{288}{25\pi}k^{-1} \approx 3.67\,k^{-1}, \qquad L_t = \frac{4\pi^3}{9\Gamma^8\left(\frac{3}{4}\right)}k^{-1} \approx 2.71\,k^{-1}. \tag{8.7}$$

Appendices 8B and 8C give details of the behavior of L_l and L_t for the ψ_b and Carter families, respectively. The next two Sections introduce different measures of focal region size, which have the advantage of providing rigorous bounds on how localized a beam can be.

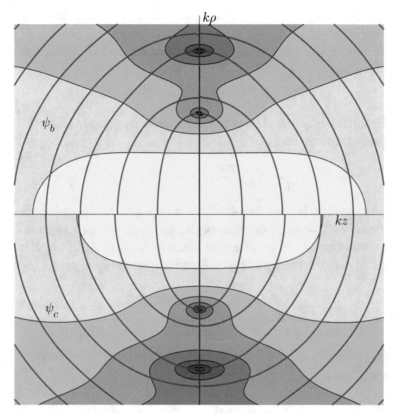

Figure 8.3: The Carter beam $\psi_C(\rho, z)$ (lower half-plane) and $\psi_b(\rho, z)$ (upper half-plane) in the focal region, plotted for $kb = 2$, for $k|z| \leq 9$, $k\rho \leq 9$. Shading indicates modulus of the wavefunction (logarithmic scale, lighter color indicates larger modulus). The isophase surfaces are shown at intervals of $\pi/2$. The phase is chosen to be zero at the origin. The isophase contours, other than those that are multiples of π, meet on the zeros of ψ, two of which are visible for both beams.

8.2   ANDREJIC MEASURE OF FOCAL PLANE INTENSITY

Andrejic (2018a,b) [1, 2] has examined the transverse localization problem from the point of view of *focal area*. He introduces a radial length s as parameter, and defines the effective focal area as

$$A(s) = \pi s^2 \frac{\int_0^\infty d\rho \, \rho \, |\psi(\rho, 0)|^2}{\int_0^\infty d\rho \, \rho \, e^{-\rho^2/s^2} |\psi(\rho, 0)|^2}. \tag{8.8}$$

The parameter s determines which part of the beam we emphasize: small s gives greater weight to the beam intensity near the beam axis, while for large s the integral in the denominator samples the intensity at large radial distances. From the area one can define an effective beam radius $\rho(s)$ by $\pi \rho^2(s) = A(s)$.

Let us see how this measure works. For the Gaussian beam fundamental mode (2.7) the expression (8.8) evaluates to

$$A_G(s) = \pi\left(s^2 + b/k\right), \qquad \rho^2(s) = s^2 + b/k. \tag{8.9}$$

As $s \to 0$ the limit $\rho(0) = \sqrt{b/k}$ is close to the value for L_t given in (8.6).

For the proto-beam we find

$$A_0(s) = \frac{\pi s^2}{1 - e^{-\sigma}[J_0(\sigma) + J_1(\sigma)]}, \qquad \sigma = \frac{(ks)^2}{2}. \tag{8.10}$$

The expansion at small s is $A_0(s) = 4\pi k^{-2} + \pi s^2 + O(s^4)$, and using the identification $\pi \rho^2(s) = A(s)$ gives $\rho(0) = 2k^{-1}$, qualitatively in agreement with L_t given in (8.7).

Andrejic shows that $4\pi k^{-2}$ is in fact the *minimum* value of $A(0)$, which is the ratio of the intensity integrated over the focal plane to the intensity at the focal center (here chosen to be the origin):

$$A(0) = \lim_{s \to 0} \frac{\pi s^2 \int_0^\infty d\rho\,\rho\,|\psi(\rho,0)|^2}{\int_0^\infty d\rho\,\rho\,e^{-\rho^2/s^2}|\psi(\rho,0)|^2} = \frac{2\pi \int_0^\infty d\rho\,\rho\,|\psi(\rho,0)|^2}{|\psi(0,0)|^2}. \tag{8.11}$$

The last equality in (8.11) results from the 'two-dimensional' delta function (for use with $\int d^2r = \int_0^{2\pi} d\phi \int_0^\infty d\rho\,\rho$)

$$\lim_{s \to 0} s^{-2}\,e^{-\rho^2/s^2} = \frac{1}{2\rho}\delta(\rho) \qquad \left(2s^{-2}\int_0^\infty d\rho\,\rho\,e^{-\rho^2/s^2} = 1\right). \tag{8.12}$$

The simplified proof that follows, of $4\pi k^{-2}$ being the minimum value of $A(0)$, assumes the wavefunction is of the form $\psi = \int_0^k d\kappa\,f(k,\kappa)\,e^{iqz} J_0(\kappa\rho)$ (Andrejic took a more general superposition). When this is so we can evaluate the numerator by using the Hankel inversion formula (3.27) which gives us the normalization of Section 2.3, namely

$$N' = \int d^2r\,|\psi|^2 = 2\pi \int_0^\infty d\rho\,\rho\,|\psi|^2 = 2\pi \int_0^k d\kappa\,\kappa^{-1}|f(k,\kappa)|^2. \tag{8.13}$$

Hence

$$A(0) = \frac{2\pi \int_0^\infty d\rho\,\rho\,|\psi(\rho,0)|^2}{|\psi(0,0)|^2} = \frac{2\pi \int_0^k d\kappa\,\kappa^{-1}|f(k,\kappa)|^2}{\left|\int_0^k d\kappa\,f(k,\kappa)\right|^2}. \tag{8.14}$$

The Andrejic proof of the minimum value of $A(0)$ is based on the Cauchy–Schwartz inequality, which we specialize to

$$\left|\int_0^k d\kappa\,F(\kappa)G(\kappa)^*\right|^2 \le \int_0^k d\kappa\,|F(\kappa)|^2 \int_0^k d\kappa'\,|G(\kappa')|^2. \tag{8.15}$$

With $F(\kappa) = f(k,\kappa)/\sqrt{\kappa}$, $G = \sqrt{\kappa}$ the inequality (8.15) reads

$$\left| \int_0^k d\kappa \, f(k,\kappa) \right|^2 \le \int_0^k d\kappa \, \kappa^{-1} \, |f(k,\kappa)|^2 \int_0^k d\kappa'\kappa' = \frac{1}{2}k^2 \int_0^k d\kappa \, \kappa^{-1} \, |f(k,\kappa)|^2. \quad (8.16)$$

The Andrejic inequality follows:

$$A(0) = \frac{2\pi \int_0^\infty d\rho \, \rho \, |\psi(\rho,0)|^2}{|\psi(0,0)|^2} = \frac{2\pi \int_0^k d\kappa \, \kappa^{-1} |f(k,\kappa)|^2}{\left| \int_0^k d\kappa \, f(k,\kappa) \right|^2} \ge 4\pi k^{-2}. \quad (8.17)$$

The proto-beam reaches this tight-focus limit, as we have seen. It is the only possible such beam, since equality in the Cauchy–Schwartz relation is attained when the functions F and G differ only by a proportionality constant. The proportionality of the functions F and G implies $f(k,\kappa) = C\kappa$, with C constant, which is the proto-beam functional form. Hence the proto-beam has the largest possible peak intensity of all physical beam wavefunctions.

For the ψ_b family we obtain, with $f(k,\kappa) = \kappa e^{qb}$ ($q = \sqrt{k^2 - \kappa^2}$),

$$A_b(0) = \frac{2\pi \int_0^k d\kappa \, \kappa \, e^{2qb}}{\left| \int_0^k d\kappa \, \kappa \, e^{qb} \right|^2} = \frac{2\pi \int_0^k dq \, q \, e^{2qb}}{\left| \int_0^k dq \, q \, e^{qb} \right|^2} = \frac{\pi b^2}{2} \frac{(2\beta - 1) e^{2\beta} + 1}{[(\beta - 1) e^\beta + 1]^2} \quad (\beta = kb). \quad (8.18)$$

The effective focal plane radius at $s = 0$ is defined by $\pi \rho^2(0) = A(0)$. The values at small and large $kb = \beta$ are

$$\rho_b(0) = \begin{cases} 2k^{-1} + \frac{1}{18}kb^2 + O(k^2b^3) \\ \sqrt{\dfrac{b}{k}} \left[1 + \dfrac{3}{4}\dfrac{1}{kb} + O((kb)^{-2}, e^{-kb}) \right]. \end{cases} \quad (8.19)$$

The small kb value $2k^{-1}$ was already obtained from (8.10). The leading large kb asymptotic term is the same as for the Gaussian beam, obtained in (8.9).

The Carter beam family has $f(k,\kappa) = \kappa e^{q^2b/2k}$. We find

$$A_C(0) = \frac{2\pi \int_0^k d\kappa \, \kappa \, e^{q^2b/k}}{\left| \int_0^k d\kappa \, \kappa \, e^{q^2b/2k} \right|^2} = \frac{2\pi \int_0^k dq \, q \, e^{q^2b/k}}{\left| \int_0^k dq \, q \, e^{q^2b/2k} \right|^2} = \frac{\pi b}{k} \frac{e^{\beta/2} + 1}{e^{\beta/2} - 1}. \quad (8.20)$$

As before we define the effective focal plane radius by $\pi\rho^2(0) = A(0)$. The values at small and large $kb = \beta$ are

$$\rho_C(0) = \begin{cases} 2k^{-1} + \dfrac{1}{48}kb^2 + O(k^3b^4) \\ \sqrt{\dfrac{b}{k}} \left[1 + 2e^{-kb/2} + O(e^{-kb}) \right]. \end{cases} \quad (8.21)$$

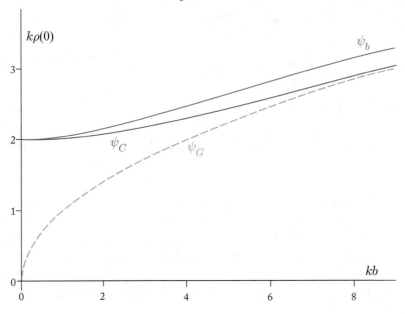

Figure 8.4: The focal region transverse extent, as measured by $\rho(0)$, defined in the text. As kb increases from zero (the proto-beam limit), the Carter beam transverse extent is smaller than that of the ψ_b family, for the same value of kb. The dashed curve shows the Gaussian beam transverse extent, $k\rho_G(0) = \sqrt{kb}$. For large kb all beam families have asymptotic transverse extent $\rho(0) \to \sqrt{b/k}$.

At small kb the value agrees with that obtained for ψ_b, since ψ_C and ψ_b have the confluent limit ψ_0. The radial extent at $kb = 0$ is $2k^{-1} = \lambda_0/\pi$, where λ_0 is the vacuum wavelength. At large kb the transverse focal extent has the same leading term $\sqrt{b/k}$ for the Carter, ψ_b and Gaussian beams. Figure 8.4 shows the effective focal radial spreads of the Carter and ψ_b beams as functions of kb.

8.3   ANDREJIC MEASURE OF LONGITUDINAL EXTENT

Andrejic (2018a,b) [1, 2] has also defined a length characterizing the longitudinal extent. In what follows we obtain the same expression by a method which is the one-dimensional version of that used in the previous section. We have a choice of definitions of longitudinal extent, for example

$$L(s) = \sqrt{\pi}\, s \frac{\int_{-\infty}^{\infty} dz\, |\psi(0,z)|^2}{\int_{-\infty}^{\infty} dz\, e^{-z^2/s^2}|\psi(0,z)|^2} \quad \text{or} \quad L(s) = 2\, s \frac{\int_{-\infty}^{\infty} dz\, |\psi(0,z)|^2}{\int_{-\infty}^{\infty} dz\, e^{-|z|/s}|\psi(0,z)|^2}. \quad (8.22)$$

In either case the value of s determines which region of the beam intensity on the beam axis is sampled preferentially: large s samples large $|z|$, small s samples small $|z|$. The factors $\sqrt{\pi}\,s$ and $2\,s$ normalize to unity for any s:

$$\frac{1}{\sqrt{\pi}s}\int_{-\infty}^{\infty} dz\, e^{-z^2/s^2} = 1, \quad \frac{1}{2s}\int_{-\infty}^{\infty} dz\, e^{-|z|/s} = 1. \tag{8.23}$$

As $s \to 0$ the sampling is of the intensity at the origin:

$$\lim_{s\,\to\,0}\ \frac{1}{\sqrt{\pi}\,s}\, e^{-z^2/s^2} = \delta(z), \quad \lim_{s\,\to\,0}\ \frac{1}{2\,s}\, e^{-|z|/s} = \delta(z). \tag{8.24}$$

In this limit both definitions in (8.22) reduce to

$$L(0) = \frac{\int_{-\infty}^{\infty} dz\, |\psi(0,z)|^2}{|\psi(0,0)|^2} = \frac{2\pi \int_0^k d\kappa\, q\kappa^{-1}|f(k,\kappa)|^2}{\left|\int_0^k d\kappa\, f(k,\kappa)\right|^2}. \tag{8.25}$$

The second equality follows from (we omit the normalization factor, which cancels)

$$\psi(0,z) = \int_0^k d\kappa f(k,\kappa)e^{iqz}, \quad \int_{-\infty}^{\infty} dz e^{i(q-q')} = 2\pi\delta(q - q') = 2\pi q\kappa^{-1}\delta(\kappa - \kappa'). \tag{8.26}$$

In the Cauchy–Schwartz inequality (8.15) we set $F(\kappa) = f(k,\kappa)\sqrt{q/\kappa}$, $G(\kappa) = \sqrt{\kappa/q}$. This gives

$$L(0) \geq 2\pi k^{-1} = \lambda_0. \tag{8.27}$$

According to the measure $L(0)$, which emphasizes the axial intensity at the center of the focal region, the least possible longitudinal extent is equal to the vacuum wavelength. Andrejic (2018b, Section 2.4 [2]) obtained the same bound by a somewhat different method.

It is interesting to examine the wavenumber weight function which gives the theoretical minimum longitudinal extent $L(0) = 2\pi k^{-1}$. Equality is attained in the Cauchy–Schwartz inequality when the functions F and G differ only by a proportionality constant. The proportionality of the functions F and G implies $f(k,\kappa) \sim \kappa q^{-1}$. But this wavenumber dependence gives a divergent beam normalization integral: from (8.13) we have $N' \sim \int_0^k d\kappa\, \kappa q^{-2} \sim \int_0^1 dy(1-y)^{-1}$, which is logarithmically divergent. Hence the theoretical minimum longitudinal extent $2\pi k^{-1} = \lambda_0$ is not attainable by beams with finite total intensity, since the intensity corresponding to $f(k,\kappa) \sim \kappa q^{-1}$ integrated over a transverse section diverges.

The proto-beam has $f(k,\kappa)$ proportional to κ, and (8.25) gives $L_0(0) = \frac{8\pi}{3}k^{-1} = \frac{4}{3}\lambda_0$. The proto-beam longitudinal extent is $4/3$ times the minimum theoretical value. For the Carter and ψ_b families the wavenumber weight functions $f(k,\kappa)$ are proportional to $\kappa e^{-b\kappa^2/2k}$, κe^{-bq}, respectively. From (8.25) we find, with $\beta = kb$ as before,

$$L_C(0) = \pi b \frac{1 - \frac{e^{-\beta}}{\sqrt{\beta}} \int_0^{\sqrt{\beta}} dt\, e^{t^2}}{\left[e^{\beta/2} - 1\right]^2} = \begin{cases} \frac{8\pi}{3} k^{-1} + \frac{4}{9}\pi b + O\left(kb^2\right) \\ \pi b - \frac{\pi}{2} k^{-1} + O\left(k^{-2}b^{-1}\right) \end{cases} \tag{8.28}$$

$$L_b(0) = \frac{\pi b}{2} \frac{\left(2\beta^2 - 2\beta + 1\right) e^{\beta} - 1}{\left[e^{\beta}\left(\beta - 1\right) + 1\right]^2} = \begin{cases} \frac{8\pi}{3} k^{-1} + \frac{4}{9}\pi b + O\left(kb^2\right) \\ \pi b + \pi k^{-1} + O\left(k^{-2}b^{-1}\right). \end{cases} \tag{8.29}$$

Note that the Carter and ψ_b families share the proto-beam limit as $kb \to 0$. The longitudinal extent of the Gaussian beam is $L_G(0) = \pi b$: on the axis the modulus squared of the Gaussian fundamental is $|\psi_G(0,z)|^2 = b^2\left(b^2 + z^2\right)^{-1}$, $\psi_G(0,0) = 1$, and $\int_{-\infty}^{\infty} dz\, |\psi_G(0,z)|^2 = \pi b$. For large kb the ψ_b and ψ_C beam families both have longitudinal extents with the leading term πb. The Carter beams are more tightly focused at the same kb value: $L_C(0) = L_b(0) - \frac{3\pi}{2}k^{-1} + O(k^{-2}b^{-1})$. Figure 8.5 shows the longitudinal extent of the focal region of the Carter and ψ_b families.

8.4   COMPARISON OF THE FOCAL EXTENT MEASURES

Sections 8.2 and 8.3 gave the Andrejic formulation of the focal extent of beams. This formulation has the advantage of providing bounds on the transverse and longitudinal localization, respectively

$$\rho(0) \geq 2k^{-1} = \lambda_0/\pi, \qquad L(0) \geq 2\pi k^{-1} = \lambda_0. \tag{8.30}$$

The transverse localization limit is reached by the proto-beam, $\psi_0(\rho, z) = 2k^{-2} \int_0^{\infty} d\kappa\, \kappa\, e^{iqz} J_0(\kappa\rho)$. The longitudinal limit cannot be reached by any beam with finite integrated intensity; it is exceeded by the factor 4/3 by the proto-beam.

The Carter and the ψ_b families of beams were compared as the delocalization parameter kb increased from zero (their common proto-beam limit). For the same value of kb the Carter beam is more localized transversely and more localized longitudinally. At large kb both families have (asymptotically) longitudinal extent πb, the Gaussian beam value. The families have a common asymptotic value of transverse extent: it is the same as the Gaussian beam value $\sqrt{b/k}$.

We emphasize that the above results, although exact, represent just one possible method of assessing beam localization. The measures proposed in Lekner (2016) [7] and discussed in Sections 2.7 and 8.1, depend on evaluating the slowly convergent integrals in the numerators of

$$\left\langle \rho^{\frac{1}{2}} \right\rangle = \frac{\int_0^{\infty} d\rho\, \rho\, \rho^{\frac{1}{2}}\, |\psi(\rho,0)|^2}{\int_0^{\infty} d\rho\, \rho\, |\psi(\rho,0)|^2} = \frac{\int_0^{\infty} d\rho\, \rho\, \rho^{\frac{1}{2}}\, |\psi(\rho,0)|^2}{\int_0^k d\kappa\, \kappa^{-1} |f(k,\kappa)|^2} \tag{8.31}$$

$$\left\langle |z|^{\frac{1}{2}} \right\rangle = \frac{\int_{-\infty}^{\infty} dz\, z^{\frac{1}{2}}\, |\psi(0,z)|^2}{\int_{-\infty}^{\infty} dz\, |\psi(0,z)|^2} = \frac{\int_{-\infty}^{\infty} dz\, z^{\frac{1}{2}}\, |\psi(0,z)|^2}{2\pi \int_0^k d\kappa\, q\, \kappa^{-1} |f(k,\kappa)|^2}. \tag{8.32}$$

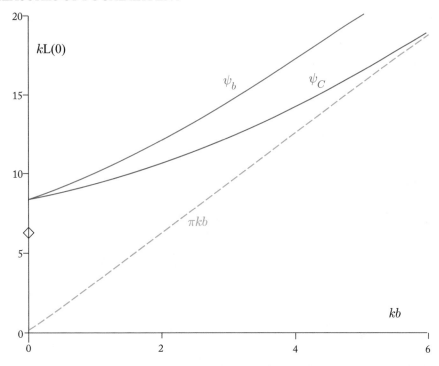

Figure 8.5: The focal region longitudinal extent for the Carter and ψ_b families, as measured by $L(0)$, defined in the text. At $kb = 0$ (the proto-beam limit) both beams have the extent $\frac{8\pi}{3}k^{-1}$, larger than the (physically unattainable) smallest possible value $2\pi k^{-1}$ (diamond). The dashed line shows the Gaussian longitudinal extent $kL_G(0) = \pi kb$, which is also the asymptotic value for large kb of both the Carter and the ψ_b families.

In contrast, the Andrejic measures in the $s \to 0$ limit have integrals which may all be evaluated in terms of the wavenumber weight function:

$$A(0) = \frac{2\pi \int_0^\infty d\rho\, \rho\, |\psi(\rho,0)|^2}{|\psi(0,0)|^2} = \frac{2\pi \int_0^k d\kappa\, \kappa^{-1} |f(k,\kappa)|^2}{\left| \int_0^k d\kappa\, f(k,\kappa) \right|^2} \tag{8.33}$$

$$L(0) = \frac{\int_{-\infty}^\infty dz\, |\psi(0,z)|^2}{|\psi(0,0)|^2} = \frac{2\pi \int_0^k d\kappa\, q\kappa^{-1} |f(k,\kappa)|^2}{\left| \int_0^k d\kappa\, f(k,\kappa) \right|^2}. \tag{8.34}$$

Comparison of (8.31) with (8.33) and of (8.32) with (8.34) shows that, in both the transverse and the longitudinal measures of beam extent, the Andrejic measures $A(0)$ and $L(0)$ emphasize the inner focal region. Of course, the more general measures $A(s)$ and $L(s)$ can give more weight

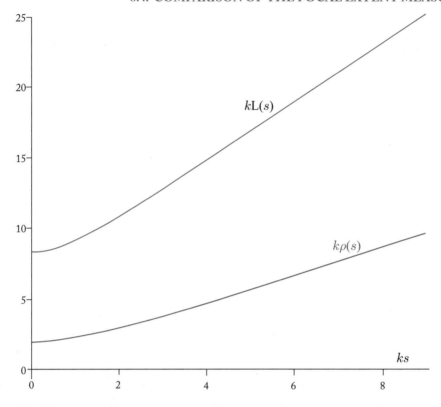

Figure 8.6: The proto-beam values of $\rho(s)$ and $L(s)$. Both increase monotonically with with s. As $s \to 0$ the limits are $\rho(0) = 2k^{-1}$ (the smallest possible value for all beams), and $L(0) = \frac{8\pi}{3}k^{-1}$, larger by $\frac{4}{3}$ than the Cauchy–Schwartz inequality suggests, but the smallest possible for a physical beam, as discussed in Section 8.3.

to outlying regions, but the simplicity and usefulness of (8.33) and (8.34) is then lost. For the proto-beam the values of $A(s)$ and $L(s)$ increase with s. Details are given in Appendix 8A.

Appendices 8B and 8C and Figure 8.7 give the longitudinal and transverse extents $\left\langle |z|^{\frac{1}{2}} \right\rangle$ and $\left\langle \rho^{\frac{1}{2}} \right\rangle$ for the ψ_b and ψ_C beam families, as function of kb. The results for the longitudinal measure are qualitatively in accord with those shown in Figure 8.6. But the transverse extent $\left\langle \rho^{\frac{1}{2}} \right\rangle$ is surprising: for both the ψ_b and ψ_C families, $\left\langle \rho^{\frac{1}{2}} \right\rangle$ initially *decreases* with kb as this parameter increases from zero. See Figure 8.7, and compare with Figure 8.4, which shows the variation of $\rho(0)$ with kb for the ψ_b and ψ_C families. A physical explanation for this unexpected behavior is currently lacking.

The results given here apply directly to quantum particle beams (Part B (Section 7.3) of Chapter 7), since in these $|\psi|^2$ is directly proportional to the probability density. Similar localization criteria might be applied to acoustic beams (Part A (Section 7.1) of Chapter 7); in these the energy density is proportional to $|\nabla\psi|^2 + k^2|\psi|^2$, and the momentum density to $\text{Im}\,(\psi^*\nabla\psi)$.

There are added difficulties for electromagnetic beams: the energy and momentum density expressions differ for different types (transverse electric, linearly polarized, circularly polarized, etc.), and depend on first and second derivatives of ψ; see Chapter 3. The tightness of focus depends on the polarization properties of the beam, with a radially polarized beam achieving localization to about $0.40\lambda_0$, whereas a linearly polarized beam achieves localization to $0.51\lambda_0$ (Quabis et al. 2000 [9], Dorn et al. 2003 [5]; the effective focal area was determined by the intensity contour at half of maximum). The sharpness of focus of higher-order radially polarized beams is discussed by Kozawa and Sato (2007) [6], and limits of the effective focal volume in multiple-beam light microscopy by Arkhipov and Schulten (2009) [3].

This chapter is based on Lekner 2020 [8], where more details can be found.

8A   APPENDIX: A(s) AND L(s) FOR THE PROTO-BEAM

We shall use the definitions (8.8) and the second of (8.22):

$$A(s) = \pi s^2 \frac{\int_0^\infty d\rho\,\rho\,|\psi(\rho,0)|^2}{\int_0^\infty d\rho\,\rho\,e^{-\rho^2/s^2}|\psi(\rho,0)|^2}, \quad L(s) = 2s\frac{\int_{-\infty}^\infty dz\,|\psi(0,z)|^2}{\int_{-\infty}^\infty dz\,e^{-|z|/s}|\psi(0,z)|^2} \tag{8A.1}$$

These have closed forms for the proto-beam $\psi_0(\rho,z)$. We give also the leading terms of $\rho(s)$ and of $L(s)$ at small and large s: from (8.10) and $\pi\rho^2(s) = A(s)$ we have

$$\rho_0(s) = \frac{s}{\{1 - e^{-\sigma}[J_0(\sigma) + J_1(\sigma)]\}^{1/2}}, \quad \sigma = \frac{1}{2}(ks)^2 \tag{8A.2}$$

$$\rho_0(s) = \begin{cases} 2k^{-1} + \dfrac{1}{4}ks^2 + O(k^3s^4) \\[2mm] s + \dfrac{1}{\sqrt{\pi}}\,k^{-1} + O(k^{-2}s^{-1}) \end{cases} \tag{8A.3}$$

$$L_0(s) = \frac{4\pi k^3 s^4}{k^2 s^2 - (1 + 3k^2 s^2)\ln(1 + k^2 s^2) + 4k^3 s^3 \arctan ks} \tag{8A.4}$$

$$L_0(s) = \begin{cases} \dfrac{8\pi}{3}k^{-1} + \dfrac{8\pi}{27}ks^2 + O(k^3 s^4) \\[2mm] 2s + \dfrac{3 + 6\ln ks}{\pi}k^{-1} + O(k^{-2}s^{-1}) \end{cases}. \tag{8A.5}$$

Figure 8.6 shows the variation of $\rho_0(s)$ and $L_0(s)$.

8B   APPENDIX: FOCAL REGION EXTENT OF THE ψ_b FAMILY

In Section 2.7 we characterized the extent of the focal region by two length: the longitudinal localization length L_l, and the transverse localization length L_t. These were defined by

$$L_l = \left\langle |z|^{\frac{1}{2}} \right\rangle^2, \qquad L_t = \left\langle \rho^{\frac{1}{2}} \right\rangle^2 \tag{8B.1}$$

$$\left\langle |z|^{\frac{1}{2}} \right\rangle = \frac{\int_0^\infty dz \, z^{\frac{1}{2}} \, |\psi(0,z)|^2}{\int_0^\infty dz \, |\psi(0,z)|^2}, \qquad \left\langle \rho^{\frac{1}{2}} \right\rangle = \frac{\int_0^\infty d\rho \, \rho \, \rho^{\frac{1}{2}} \, |\psi(\rho,0)|^2}{\int_0^\infty d\rho \, \rho \, |\psi(\rho,0)|^2}. \tag{8B.2}$$

We shall look at the *longitudinal extent* first. From (2.37) we know that

$$\begin{aligned}
\psi_b(0,z) &= \frac{b^2}{\left[e^{kb}(kb-1)+1 \right]} \int_0^k dq \, q \, e^{q(b+iz)} \\
&= \left(\frac{b}{b+iz} \right)^2 \frac{e^{k(b+iz)} \left[k(b+iz)-1 \right]+1}{\left[e^{kb}(kb-1)+1 \right]}
\end{aligned} \tag{8B.3}$$

The normalization integral evaluates (with $kb = \beta$ as before) to

$$k \int_0^\infty dz \, |\psi_b(0,z)|^2 = \frac{\pi \beta}{2} \frac{\left[\left(\beta^2 - \beta + \frac{1}{2} \right) e^{2\beta} - \frac{1}{2} \right]}{\left[(\beta-1) e^\beta + 1 \right]^2}. \tag{8B.4}$$

The average value of $|z|^{\frac{1}{2}}$ involves error functions of both real and imaginary argument:

$$\left\langle |kz|^{\frac{1}{2}} \right\rangle = \sqrt{2\beta} \frac{4\pi \left(\beta^2 - \frac{\beta}{2} + \frac{1}{4} \right) e^{2\beta} \mathrm{erf}\sqrt{\beta} + 4\sqrt{\pi\beta} e^\beta (\beta-1) - i\pi \mathrm{erf}\, i\sqrt{\beta}}{4\pi \left[\left(\beta^2 - \beta + \frac{1}{2} \right) e^{2\beta} - \frac{1}{2} \right]}. \tag{8B.5}$$

As $\beta \to 0$ we regain the value $L_l = \left\langle |z|^{\frac{1}{2}} \right\rangle^2 \to 288/25\pi k$ of Equation (2.59). For large β the focal longitudinal extent is proportional to b, as expected:

$$L_l = 2b + 2/k + O(1/k^2 b) = 2b \left[1 + \beta^{-1} + O(\beta^{-2}) \right]. \tag{8B.6}$$

The *transverse extent* is more difficult to evaluate. We need integrals over the focal plane wavefunction (un-normalized)

$$\psi_b(\rho,0) = F_b(\rho) = \int_0^k dq \, q \, e^{qb} J_0 \left(\rho\sqrt{k^2 - q^2} \right). \tag{8B.7}$$

The expectation value to be evaluated is

$$\left\langle \rho^{\frac{1}{2}} \right\rangle = \frac{\int_0^\infty d\rho \, \rho \, \rho^{\frac{1}{2}} \, F_b^2(\rho)}{\int_0^\infty d\rho \, \rho \, F_b^2(\rho)} \equiv \frac{N}{D}. \tag{8B.8}$$

The denominator in the expectation value of $\sqrt{\rho}$ can be evaluated by means of the Hankel inversion formula (4B.7):

$$
\begin{aligned}
D &= \int_0^\infty d\rho \, \rho \, F_b^2(\rho) = \int_0^\infty d\rho \, \rho \int_0^k d\kappa \, \kappa \, e^{qb} \, J_0(\kappa\rho) \int_0^k d\kappa' \, \kappa' \, e^{q'b} \, J_0(\kappa'\rho) \\
&= \int_0^k d\kappa \kappa \, e^{qb} \int_0^k d\kappa' \, \kappa' \, e^{q'b} \, \kappa^{-1} \delta\left(\kappa - \kappa'\right) = \int_0^k d\kappa \, \kappa \, e^{2qb} \qquad\text{(8B.9)} \\
&= \int_0^k dq \, q \, e^{2qb} = \frac{1}{4b^2}\left[(2\beta - 1)\, e^{2\beta} + 1\right].
\end{aligned}
$$

The numerator $N = \int_0^\infty d\rho \, \rho^{\frac{3}{2}} \, F_b^2(\rho)$ is slowly convergent. For the wavefunction in the focal plane we use the results (2B.12) and (2B.25) for the proto-beam wavefunction $\psi_0\,(\rho, z) = F + iG$, namely

$$
G(\rho, z) = \frac{2kz}{(kr)^3}[\sin kr - kr\cos kr] = \frac{2z}{r}j_1(kr) \qquad\text{(8B.10)}
$$

$$
F(\rho, z) = 2\sum_0^\infty \frac{(-)^n 2^n n!(kz)^{2n}}{(2n)!}\frac{J_{n+1}(k\rho)}{(k\rho)^{n+1}}. \qquad\text{(8B.11)}
$$

To obtain $\psi_b\,(\rho, 0)$ we set $z \to z - ib$ in both, and then let $z \to 0$. We also omit the prefactor 2 in F and G because we are using an un-normalized wavefunction (8B.7). The result is, with $R = \sqrt{\rho^2 - b^2}$,

$$
\psi_b(\rho, 0) = F_b(\rho) = \frac{b}{R}j_1(kR) + \sum_0^\infty \frac{2^n n!(kb)^{2n}}{(2n)!}\frac{J_{n+1}(k\rho)}{(k\rho)^{n+1}}. \qquad\text{(8B.12)}
$$

Since $n! \to \sqrt{2\pi n}(n/e)^n$ for large n, the number of terms required in the series is of order β. The term with longest range in ρ is the $n = 0$ term of the series, $\psi_0 = (k\rho)^{-1}J_1(k\rho)$. We write

$$
F_b(\rho) = \psi_0 + \psi_1, \quad \psi_1 = \frac{b}{R}j_1(kR) + \sum_1^\infty \frac{2^n n!(kb)^{2n}}{(2n)!}\frac{J_{n+1}(k\rho)}{(k\rho)^{n+1}}. \qquad\text{(8B.13)}
$$

The numerator we rewrite as $N = \int_0^\infty d\rho \, \rho^{\frac{3}{2}}\left[\psi_0^2 + 2\psi_0\psi_1 + \psi_1^2\right]$, and evaluate the integral over ψ_0^2 as in Section 2.7 to regain the $kb \to 0$ value $L_t = \left\langle\rho^{\frac{1}{2}}\right\rangle^2 \to 4\pi^3/(9\Gamma^8\left(\frac{3}{4}\right)k)$ of Equation (2.59). The remaining integrations converge rapidly enough to give the numerical values plotted in Figure 8.7. The result is surprising: the transverse extent of the ψ_b family at first decreases with increasing kb. The same behavior is shown by the Carter beams defined in (2.26). These share the proto-beam limit with the ψ_b family as $kb \to 0$. The details are given in Appendix 8C; the resulting longitudinal and transverse extents are shown here in Figure 8.7.

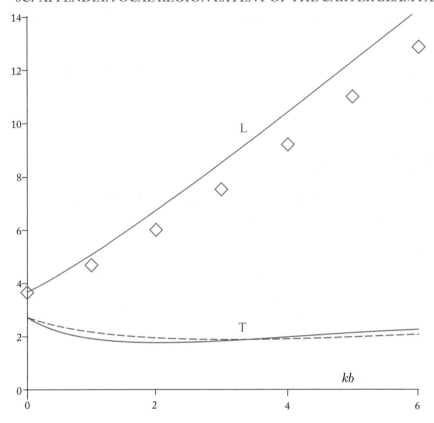

Figure 8.7: Longitudinal and transverse extent of the focal region of scalar ψ_b beams, as functions of $\beta = kb$. The curves labeled L and T show $k\left\langle |z|^{\frac{1}{2}} \right\rangle^2$ and $k\left\langle \rho^{\frac{1}{2}} \right\rangle^2$, respectively. The dashed curve and points are for the Carter beam family, discussed in Appendix 8C.

8C   APPENDIX: FOCAL REGION EXTENT OF THE CARTER BEAM FAMILY

We wish to calculate the longitudinal and transverse extents of the family defined in (2.26):

$$\psi_C\left(\rho,\, z\right) = \frac{b/k}{1 - e^{-kb/2}} \int_0^k d\kappa\, \kappa\, e^{-b\kappa^2/2k + iqz}\, J_0\left(\kappa\rho\right)$$

$$= \frac{b/k}{e^{kb/2} - 1} \int_0^k dq\, q\, e^{bq^2/2k + iqz}\, J_0\left(\kappa\rho\right). \tag{8C.1}$$

The prefactor to the integral is chosen to normalize the beam wavefunction to unity at the origin.

On the beam axis we find (again setting $kb = \beta$)

$$\psi_C(0, z) = \frac{e^{\beta/2+ikz} - 1}{e^{\beta/2} - 1} - \sqrt{\frac{\pi}{2\beta}} \frac{kze^{kz^2/2\beta}\left[\text{erf}\left(\frac{kz}{\sqrt{2\beta}}\right) - \text{erf}\left(\frac{kz - i\beta}{\sqrt{2\beta}}\right)\right]}{e^{\beta/2} - 1}. \quad (8C.2)$$

The asymptotic form of $|\psi_C(0, z)|^2$ is

$$|\psi_C(0, z)|^2 = \frac{\beta^2 e^\beta}{(e^{\beta/2} - 1)^2}\left\{z^{-2} - 2z^{-3}e^{-\frac{\beta}{2}}\sin kz + O(z^{-4})\right\}. \quad (8C.3)$$

It is necessary to go to high order in z^{-1} for good accuracy in the evaluation of $\langle|z|^{\frac{1}{2}}\rangle$ defined in (8B.2), and to choose domains $0 \le z \le a$, $a \le z < \infty$ in the integrations and sum the results. The values chosen in calculation of the points in Figure 8.6 were $ka = kb + 5$.

Next we look at the transverse extent of the Carter family of beams. The wavefunction in the focal plane $z = 0$ is

$$\psi_C(\rho, 0) = \frac{b/k}{1 - e^{-kb/2}} \int_0^k d\kappa\, \kappa\, e^{-b\kappa^2/2k} J_0(\kappa\rho). \quad (8C.4)$$

The asymptotic form has the leading terms

$$\psi_C(\rho, 0) \sim \frac{\beta}{e^{\beta/2} - 1}\left[\frac{J_1(k\rho)}{k\rho} - \beta\frac{J_0(k\rho)}{(k\rho)^2}\right]. \quad (8C.5)$$

In the denominator of the $\langle\rho^{\frac{1}{2}}\rangle$ integral (8B.8) we write

$$\begin{aligned}
D &= \int_0^\infty d\rho\, \rho\, \psi_C(\rho, 0)^2 \\
&= \left(\frac{b/k}{1 - e^{-kb/2}}\right)^2 \int_0^\infty d\rho\, \rho \int_0^k d\kappa\, \kappa\, e^{-b\kappa^2/2k} J_0(\kappa\rho) \int_0^k d\kappa'\, \kappa'\, e^{-b\kappa'^2/2k} J_0(\kappa'\rho) \\
&= \left(\frac{b/k}{1 - e^{-kb/2}}\right)^2 \int_0^k d\kappa\, \kappa\, e^{-b\kappa^2/2k} \int_0^k d\kappa'\, \kappa'\, e^{-b\kappa'^2/2k} \kappa^{-1}\delta(\kappa - \kappa') \\
&= \left(\frac{b/k}{1 - e^{-kb/2}}\right)^2 \int_0^k d\kappa\, \kappa\, e^{-b\kappa^2/k} = \frac{\beta}{2}\frac{e^{\beta/2} + 1}{e^{\beta/2} - 1}.
\end{aligned} \quad (8C.6)$$

To evaluate the numerator $N = \int_0^\infty d\rho\, \rho^{3/2}\, \psi_C(\rho, 0)^2$ we split the integration into the domains $0 \ldots a$, $a \ldots \infty$, and use the asymptotic form (8B.5) for the latter. The value $ka = 7.5$ was used to generate the dashed curve in Figure 8.7.

The results for the longitudinal and transverse extents of the Carter beam family are shown in Figure 8.7. They agree exactly with those of the ψ_b family as $\beta = kb \to 0$, and closely thereafter.

8.8   CITED REFERENCES

[1] Andrejic, P. 2018a. Convergent measure of focal extent, and largest peak intensity for non-paraxial beams, *Journal of Optics*, 20(075610):10. DOI: 10.1088/2040-8986/aaca6b. 147, 150, 153

[2] Andrejic, P. 2018b. Localised waves, Master of Science Thesis, Victoria University of Wellington. 147, 150, 153, 154

[3] Arkhipov, A. and Schulten, K. 2009. Limits for reduction of effective focal volume in multiple-beam light microscopy, *Optics Express*, 17:2861–2870. DOI: 10.1364/oe.17.002861. 158

[4] Carter, W. H. 1973. Anomalies in the field of a Gaussian beam near focus, *Optics Communications*, 64:491–495. DOI: 10.1016/0030-4018(73)90012-6. 147

[5] Dorn, R., Quabis, S., and Leuchs, G. 2003. Sharper focus for a radially polarized light beam, *Physics Review Letters*, 91(233901):1–4. DOI: 10.1103/physrevlett.91.233901. 158

[6] Kozawa, Y. and Sato, S. 2007. Sharper focal spot formed by higher-order radially polarized laser beams, *Journal of the Optical Society of America*, 24:1793–1798. DOI: 10.1364/josaa.24.001793. 158

[7] Lekner, J. 2016. Tight focusing of light beams: A set of exact solutions, *Proc. of the Royal Society A*, 472(20160538):17. DOI: 10.1098/rspa.2016.0538. 147, 155

[8] Lekner, J. 2020. Focal extent of scalar beams, *J. Opt.*, 22:045607. 158

[9] Quabis, S., Dorn, R., Eberler, M., Glöckl, O., and Leuchs, G. 2000. Focusing light to a tighter spot, *Optics Communications*, 179:1–7. DOI: 10.1016/s0030-4018(99)00729-4. 158

8.9   ADDITIONAL REFERENCES (CHRONOLOGICAL ORDER)

[10] Porras, M. A. 1994. The best quality optical beam beyond the paraxial approximation *Optics Communications*, 111:338–349. DOI: 10.1016/0030-4018(94)90475-8.

[11] Berry, M. V. 1998. Wave dislocation reactions in non-paraxial Gaussian beams, *Journal of Modern Optics*, 45:1845–1858. DOI: 10.1080/09500349808231706.

[12] Nye, J. F. 1998. Unfolding of higher-order wave dislocations, *Journal of the Optics Society of America*, 15:1132–1138. DOI: 10.1364/josaa.15.001132.

[13] Lekner, J. 2004a. Invariants of atom beams, *Journal of Physics B: Atomic and Molecular Physics*, 7:1725–1736. DOI: 10.1088/0953-4075/37/8/013.

[14] Lekner, J. 2004b. Invariants of three kinds of generalized Bessel beams, *Journal of Optics A: Pure Applied Optics*, 6:837–843. DOI: 10.1088/1464-4258/6/9/004.

[15] Lekner, J. 2006. Acoustic beams with angular momentum, *Journal of the Acoustic Society of America*, 120:3475–3478. DOI: 10.1121/1.2360420.

[16] Zamboni-Rached, M. 2009. Unidirectional decomposition method for obtaining exact localized wave solutions totally free of backward components, *Physics Review A*, 79(013816). DOI: 10.1103/physreva.79.013816.

[17] Garay-Avendaño, R. L. and Zamboni-Rached, M. 2014. Exact analytic solutions of Maxwell's equations describing nonparaxial electromagnetic beams, *Applied Optics*, 53:4524–4531. DOI: 10.1364/AO.53.004524.

[18] Zamboni-Rached, M., Ambrosio, L. A., Dorrah, A. H., and Mojahedi, M. 2017. Structuring light under different polarization states within micrometer domains: Exact analysis from the Maxwell equations, *Optics Express*, 25:10051–10056. DOI: 10.1364/oe.25.010051.

[19] Andrejic, P. and Lekner, J. 2017. Topology of phase and polarization singularities in focal regions, *Journal of Optics*, 19(105609):8. DOI: 10.1088/2040-8986/aa895d.

[20] Philbin, T. G. 2018. Some exact solutions for light beams, *Journal of Optics*, 20(105603). DOI: 10.1088/2040-8986/aade6d.

[21] Lekner, J. and Andrejic, P. 2018. Nonexistence of exact solutions agreeing with the Gaussian beam on the beam axis or in the focal plane, *Optics Communications*, 407:22–26. DOI: 10.1016/j.optcom.2017.08.071.

Author's Biography

JOHN LEKNER

John Lekner is Emeritus Professor of theoretical physics at Victoria University of Wellington, New Zealand. After an M.Sc. at the University of Auckland and Ph.D. at the University of Chicago, he taught at the Cavendish Laboratory, Cambridge, where he was also Fellow and Tutor in Physics at Emmanuel College. He has worked in statistical physics, electromagnetism, quantum theory and theory of fluids. He is the author of 157 papers and of the books *Theory of Reflection* (2ed, Springer 2016) and *Theory of Electromagnetic Pulses* (IOP Concise Physics, 2018).

Printed in the United States
by Baker & Taylor Publisher Services